The Spiritual Dimension of Business Ethics and Sustainability Management

László Zsolnai
Editor

The Spiritual Dimension of Business Ethics and Sustainability Management

Springer

Editor
László Zsolnai
Corvinus University of Budapest
Hungary

ISBN 978-3-319-11676-1 ISBN 978-3-319-11677-8 (eBook)
DOI 10.1007/978-3-319-11677-8
Springer Cham Heidelberg New York Dordrecht London

Library of Congress Control Number: 2014956383

Printed on acid-free paper

Springer is part of Springer Science+Business Media (www.springer.com)

Preface

This volume introduces the dimension of spirituality into business ethics and sustainability management. It presents selected papers from the international "*Spirituality and Sustainability: A New Path for Entrepreneurship*" conference held in September 21–23, 2012 in Visegrad, Hungary. The conference was organized by the Business Ethics Center of Corvinus University of Budapest in cooperation with the European SPES Forum and ERENET—the Entrepreneurship Research and Education Network of Central European Universities.

The volume presents and summarizes new perspectives and findings about sustainability-related spirituality from a social and economic perspective. It reports on innovative practices and policy reform and discusses spiritual-based leadership for sustainability management. The main function of the volume is to present ideas and initiatives that can lead toward responsible business practices and promote policies for ecological sustainability. It seeks a value community of readers: scholars, practitioners and policymakers who are engaged with genuine ethics in business, environmental management and public policy.

The volume is a product of 19 scholars and practitioners from Europe, North America and Asia. The contributors represent a diversity of fields, including business ethics, management, finance, leadership, religious studies, literature, and sociology.

It is the hope of the editor that the papers presented in this volume will be good food for thought and inspiration to action for those who are engaged in the never-ending "business" of preserving and restoring nature.

June 2014
Budapest

László Zsolnai

Contents

Part I
Introduction

Spirituality, Ethics and Sustainability

Laszlo Zsolnai

Martin Heidegger claimed that modern technology, with its violent metaphysics, destroys Being (Heidegger 1978). However, it is not modern technology itself that is dangerous but its careless and wide-scale implementation by modern-day business. With its exclusive focus on profit-making, modern-day businesses tend to violate the integrity and diversity of natural ecosystems, the autonomy and culture of local communities and the chance that future generations will lead a decent life.

The metaphysics of modern-day business can be described by the following statements: (i) 'to be' is to be a marketable resource; (ii) 'to be' involves being either an object available for productive activity on the market, or else a subject who makes use of such objects; and, (iii) the only mode of thinking is calculative thinking; the consideration and measurement of every being as a marketable resource (Young 2002).

Such *market metaphysics*—what George Soros (1998) rightly calls "market fundamentalism"—necessarily lead to the violation of natural and human beings. In many cases violent business practices result in 'essential' harms such as the exploitation of forests for timber or the commoditization of women as mere sex objects.

To preserve nature and to satisfy real human needs, gentle, careful ways of undertaking economic activities are needed. This requires that economic actors have the intrinsic motivation to serve the greater good, and are ready to measure success using broader value categories than money alone.

The central claim of this book is that both *business ethics* and *sustainability management* require *spirituality* as a *foundation*. Without spiritually-motivated actors, ethical business initiatives and pro-environmental activities can become ineffective and meaningless, and sometimes even counter-productive and destructive.

L. Zsolnai (✉)
Corvinus University of Budapest, Budapest, Hungary
e-mail: laszlo_zsolnai@interware.hu

L. Zsolnai (ed.), *The Spiritual Dimension of Business Ethics and Sustainability Management*, DOI 10.1007/978-3-319-11677-8_1

Both "spirituality" and "sustainability" are complex notions, hence a definition of the terms may help to set the context in which we can discuss the relationship between spirituality, business ethics and sustainability management.

Spirituality cannot be captured in one standardized definition. Spirituality is a rich, intercultural and multilayered concept. As a guideline I propose to use the following working definition (developed by the European SPES Forum): *spirituality is people's multiform search for the deep meaning of life that interconnects them to all living beings and to 'God' or Ultimate Reality* (European SPES Forum 2014).

Most definitions of spirituality share a number of common elements: reconnection to the inner self; a search for universal values that lifts the individual above egocentric strivings; deep empathy with all living beings; and, finally, a desire to somehow keep in touch with the source of life (whatever name we choose to give this). In other words, spirituality is the search for inner identity, connectedness and transcendence (Bouckaert and Zsolnai 2012).

While the notion of spirituality was for a long time exclusively related to religion, today it clearly goes beyond the boundaries of institutional religions. For believers, and nonbelievers as well, spirituality can function:

1. as a transconfessional good and thereby a suitable platform for interreligious dialogue that can help move humanity beyond clashes of religions and cultures;
2. as a public good, not only a private matter—which demands an appropriate form of public management;
3. as a profane good that does not isolate the spiritual but rather integrates it into political, social, economic and scientific activities;
4. as an experience-based good that is accessible to each human being through their reflections on their own inner life;
5. as a source of inspiration in the human and social quest for meaning (Bouckaert and Zsolnai 2012).

Now, let us turn to the meaning of *sustainability* in relation to business. Business affects the natural environment at different levels of the organization of nature:

I. Individual organisms are affected by business via hunting, fishing, agriculture, experimentation on animals, etc.
II. Natural ecosystems are affected by business via mining, the regulation of rivers, building, the pollution of air, water and land, etc.
III. The Earth as a whole is affected by business via the extermination of species, the impacts of climate change, etc. (Zsolnai 2011).

Sustainability can be interpreted to mean *non-declining natural wealth*. Environmental ethics—developed by Jonas (1984), Fox (1990), Singer (1995), Leopold (1949), Lovelock (2000) and others—offer well-established operating principles that business can follow to move towards sustainability:

1. Business should promote natural living conditions and a pain-free existence for animals and other sentient beings.
2. Business should use natural ecosystems in such a way that their health is not damaged during use.

3. Business should not contribute to violating the systemic patterns and global operating mechanisms of the Earth (Zsolnai 2011).

In summary, we can say that business is sustainable if and only if its aggregate impact on animal welfare, ecosystem health and the living Earth is non-negative.

Current experience shows us that, with its materialistic management paradigm, mainstream business is not sustainable according to the above-mentioned parameters (Tencati and Zsolnai 2010).

The *materialistic management paradigm* is based on the belief that the sole motivation behind doing business is money-making, and success should be measured only in terms of the profit that is generated. However, new values are emerging in modern business that contribute to a *post-materialistic management paradigm*: frugality, deep ecology, trust, reciprocity, responsibility for future generations, and authenticity. Within this framework making profits and creating growth are no longer ultimate aims but elements of a wider set of values. In a similar way, cost-benefit calculations are no longer considered the essence of management but rather part of a broader concept of wisdom in leadership. Spirit-driven businesses employ their intrinsic motivation to serve the common good and use holistic evaluation schemes to measure their success (Bouckaert and Zsolnai 2011).

The present volume is dedicated to elaborating the emerging post-materialistic business paradigm by focusing on the spiritual dimensions of business ethics and sustainability management.

After this introductory chapter *Part 2* of the book presents *new perspectives and findings* in business ethics and sustainability management through using a spiritual perspective.

In his paper, "Spirituality: The Missing Link in Business Ethics", business ethicist *Luk Bouckaert* (Catholic University of Leuven, Belgium) argues that a more *spiritual approach* to *business ethics* is needed. Without the involvement of intrinsic motivation business ethics is reduced to being an instrument of reputation and risk management and any genuine moral commitment is lost. Bouckaert stresses that, if we need to reshape our economic, political and religious institutions, we need something that can restore a sense of shared meaning, responsibility and purpose. Spirituality should therefore be promoted as a public good and a public virtue. This approach should be linked to the practice of spiritually-based leadership and to a deep sense of social responsibility.

In his paper, "Spiritual Sustainability Management", philosopher *Hendrik Opdebeeck* (University of Antwerp, Belgium) claims that in exploring a new path for entrepreneurship one should bridge management and wisdom. In order to achieve this, he suggests to explore the richness of Islamic and Jewish traditions.

Opdebeeck emphasizes that two great cultures have transferred the ancient traditions of Europe to the Christian West: Arabic and Jewish culture. Religious Judaism developed the foundations on which later Christianity and Western Christian culture were built. Islam regarded the Jewish and Christian religions as precursors of Islam and therefore preached a policy of respect and tolerance for the Jewish and Christian minorities in their midst. One of the most important Jewish philosophers between Antiquity and the Renaissance era was Moses Maimonides (1135–1206), who transmitted the Aristotelian tradition of practical wisdom. Its relevance to sustainable entrepreneurship today is that Maimonides and other Jewish philosophers like Gabirol, along with Arab thinkers like Avicenna and Averroes, not only preserved the Greek tradition but also accurately and fruitfully considered how one may create a bridge between rational thought and spiritual tradition.

In his paper, "The Inner Perspective—the Sufi Approach", management consultant *András László* (EuroVisioning, Brussels, Belgium) gives a concise introduction to Sufism. The teachings of Sufi Masters Hazrat and Pir Vilayat Inayat Khan set the tone and framework within which the understanding of the inner perspective on sustainability and leadership can gradually unfold. The paper highlights the new settings/dimensions of consciousness which may be identified according to Sufism; the transcendent dimension (awakening from personal consciousness), the spiritual dimensions of emotions and creativity (tapping in to the thinking of the Universe), and the shifting of perspective (turning within, and the search for meaning).

Within this Sufi framework, the paper explores the much-needed "LeaderShift" in our age of mass collaboration and ever-deepening interconnectivity. It puts at the forefront the recovery of the Soul, the Heart and Wisdom as pivotal added value.

In his paper, "Deep Ecology and Personal Responsibility", business ethicist *Knut J. Ims* (Norwegian School of Economics, Bergen, Norway) explores the deep ecology approach.

According to Norwegian philosopher Arne Naess, deep ecology has two goals: (1) try to bring about a change in attitudes to nature; and, (2) develop an alternative philosophy called "eco-philosophy". While eco-philosophy is a descriptive science, ecosophy also includes the study of the development and eliciting of norms. Arne Næss invites everybody to work out his or her individual ecosophy, by which he means one's personal philosophy of ecological harmony. This kind of wisdom, is openly normative, and contains value priorities, as well as norms. In his ecosophy Naess negates the dominant metaphysics of the modern age which would have humans to be essentially different from the rest of nature.

In his paper "Spiritual Orientation to Nature in Buddhism", Buddhist scholar *Gábor Kovács* (Corvinus University of Budapest, Hungary) stresses that the Western anthropocentric worldview has separated human beings from nature and made them the dominant species on the planet. This worldview has resulted in a massive environmental crisis. Modern societies are detached from nature and manage their businesses in unsustainable ways. The consumption-based human lifestyle is depleting the resources of the planet at an alarming rate and polluting the environment on a large-scale.

Buddhism refutes the dominance of humans over other species and avoids taking an anthropocentric worldview. The heart of Buddhism is the quest for spiritual perfection through practice. The observation of precepts and the development of virtues are the first steps towards creating a non-harming and peaceful lifestyle. Meditation and the calming of the mind result in further strengthening of virtues and precepts. Continuous practice may create a positive spiral which may produce spiritual perfection and a sustainable life-style. Kovács emphasizes that sustainability in Buddhism is built around the conception of interdependence and the practice of non-harming.

In her paper "Sustainability and Wisdom: The Power of the Fable" literary scholar *Rita Ghesquière* (Catholic University of Leuven, Belgium) analyses the fable as an old literary genre/model that has its roots in popular tradition. In the Western tradition, Aesop was one of the first to give the fable literary prestige. Phaedrus and Jean de La Fontaine followed in his footsteps. The fable, like the fairy tale, can be thought of as a global, common good. We find fables in cultures from around the world and often they have similar messages of wisdom.

Ghesquière analyses fables and extracts examples of sustainable economic behavior. In Aesop famous fables—"The cricket and the ant", "The farmer and his sons", "The fox with the big belly", and "The miser and his gold"—sustainability is linked with wisdom, frugality, and having a vision of the long term.

Part 3 of the book introduces *innovative practices* and *proposals for reforms* for sustainability.

In his paper, "Sustainocracy: Spirituality and Sustainable Progress", environmental activist and community organizer *Jean-Paul Close* (STIR Foundation, AiREAS and STIR Academy, The Netherlands) argues that all the crises, recessions and related chaos that we suffer today are a consequence of the breaking up of old fragmented system hierarchies to make way for new, local, modern, holistic and cooperative formats of human interaction based on equality, trust and a high level of spiritual awareness.

Close started establishing unique cooperative formats ('AiREAS') in The Netherlands in which government, business, science and environmental authorities take joint responsibility for a complex, human objective (e.g. assuring air quality and health, multiculturality and social cohesion). As old paradigms crumble in the shape of further recessions, new ones receive room to grow, eventually even producing a quantum leap in human evolutionary progress.

In his paper, "Quakers and Climate Change", environmental scholar and social activist *Laurie Michaelis* (Society of Friends, Oxford, UK) presents the *Living Witness Project* of British Quakers which supports a number of Quaker groups in their experiments with collaborative approaches to sustainable living. Michaelis shares some of the experiences and insights arising from that project and describes the distinctive Quaker approach to power, truth and inclusion in a sustainability context.

In his paper, "Sustainability and Long Term Growth in the Financial Market System", finance expert *Aloy Soppe* (Erasmus University of Rotterdam, The Netherlands) argues that capital global growth rates should be similar to rates of real economic growth. As an extension of the social discount rate, extreme interest rates (either too low or too high) may be interpreted as premiums because of the assumed sustainability-related premise that long-term financial claims should equal long-term physical assets. From that perspective, interest rates do not just reflect the market price of capital but are variables with important implications for social welfare and the sustainability of the economy.

In their paper "Spiritual-Based Entrepreneurship for an Alternative Food Culture: The Transformational Power of Navdanya" business consultant *Arundhati Virmani* (EHESS-CNRS, Marseille and ESC Rennes School of Business, France) and business ethicist *François Lépineux* (ESC Rennes School of Business, France) examine the mobilization of spiritual notions in defense of sustainable agriculture in India through the specific case of Navdanya, a network of seed keepers and organic producers.

Navdanya has successfully used the spiritual language of protest from its national anti-colonial traditions and battles to safeguard and revive Indian rural and artisan activities. Navdanya's fight for rights and common good is articulated in a language rooted in the traditional wisdom and knowledge of common folk. Navdanya offers an example of how spiritual values and traditions can mobilize people to engage in sustainable agriculture and can defend them against misuses of corporate power.

In his paper, "Enforcement of Self-Interests by Nature Transformers", ecologist and environmental activist *Janos Vargha* (Danube Circle, Budapest, Hungary) shows how large-scale technological projects—particularly hydroelectric dams—destroy natural ecosystems. He demonstrates that planners and builders manipulate politics and the general public by misrepresenting and disregarding the negative environmental effects of their projects. Vargha suggests employing a saying attributed to Comenius—"Let everything flow free, violence shall be afar from things"—as a general guiding principle for judging the ecological and ethical merits of technological projects.

In her paper "Entrepreneurship Inspired by Nature" environmental management scholar *Nel Hofstra* (Erasmus University Rotterdam, The Netherlands) states that our perspectives about nature determine how we evaluate environmental problems, how we seek to conserve nature and try to prevent future problems. Nature has been over-exploited by mainstream business. To constrain further destruction, a shift from an anthropocentric to a more eco-centric approach is developing entrepreneurship and eco-innovations, expanding the body of knowledge about Nature's operating instructions.

Part 4 of the book presents a debate about spiritual-based leadership in business.

In his lead paper, "Spiritual-based Leadership", business ethicist *Peter Pruzan* (Copenhagen Business School, Denmark & Sri Sathya Sai Institute of Higher Learning, Prasanthi Nilayam, India) argues that the concept of spiritual-based leadership is emerging as approach to leadership that integrates a leader's inner perspectives on identity, purpose, responsibility and success with his or her decisions and actions in the outer world of business—and it is also emerging as a significant framework for understanding, practicing, communicating and teaching the art and profession of leadership.

Pruzan notes that spiritual-based leaders are nourished by their spirituality, which is a source that informs and guides them. They search for meaning, purpose and fulfillment in the external world of business and in the internal world of consciousness and conscience. Their external actions and their internal reflections are mutually supportive, so that rationality and spirituality are also mutually supportive.

In her article, "Educating Moral Business Leaders Without the Fluff and Fuzz", leadership scholar *Joanne Ciulla* (University of Richmond, USA) argues that spiritual-based leadership is only one of many ways to describe how to develop and educate moral leaders. We want business leaders who do good things. Whether they are motivated by spirituality, humanism, virtue or an enhanced intellectual perspective is not as important as whether they behave ethically and sustainably. Educators and scholars must seek the most effective ways to educate business students to take on the moral responsibilities of leadership.

While Ciulla is sympathetic to the ideals of spiritual-based leadership, she doubts that promoting spirituality is the best way to convince business schools to change how or what they teach. The task of educating moral business leaders is too important to approach without having clear and compelling ideas about what moral leadership is and how to practice it. If we get rid of the fuzz, people may stop calling ethics a fluffy subject.

In his article, "In Admiration of Peter Pruzan's proposal for Spiritual-based Leadership", business management expert *Steven B. Young* (Caux Round Table, Minneapolis, USA) notes that good leadership resonates with vision and mission and looking beyond the present. Young emphasizes that we are both material and spiritual. To insist on one and not the other is to deny the truth of our humanity. The Spirit must enter the world through our action, and action may be demonstrated in leadership.

In his article, "The Paradox of Pragmatism", business spirituality scholar *Paul de Blot, SJ* (Nyenrode Business University, Breukelen, The Netherlands) suggests that the dualism of 'the inside' and 'the outside' cannot be solved if we start by focusing on the differences. We should start by examining unity. He refers to Nobel prize winning chemist Ilya Prigogine considering an open system with a chaotic outside which developed internal harmony. Jesuit biologist Teilhard de Chardin saw evolution as being self conscientization of the material world, also developing from a spiritual inside. French philosopher Gabriel Marcel looked for meaningfulness in these relationships by reflecting on modes of having and being.

In her article, "Cultivating Moral Humility", management scholar *Katalin Illés* (University of Westminster, London, UK) notes that great leaders quite naturally make the connection between their external and internal worlds and relate to the environment in a humble and responsible manner, expressing their concerns for the common good as something deeply personal.

Illés believes that the only way to contribute to changing humanity is to start changing ourselves. The wisdom literature from different cultural traditions tells us to pay attention to our inner world, to search for answers in the silence of our hearts, to sit with our dilemmas and patiently wait for answers, to listen to the inner wisdom that we all carry inside.

The debate ends with a replay by *Peter Pruzan*, "Further Reflections on Spirituality and Spiritual-based leadership". He emphasizes the fact that unless a leader's behavior is grounded on existential inquiry that leads to self-knowledge, ethics, no matter how it is taught or what ethical codices are developed to guide behavior in organizations, will, continue to succumb to the demands of economic rationality. Scandals will continue to anger and frustrate us no matter how much focus our business schools place on business ethics and how many Green Papers are developed to promote corporate social responsibility.

The knowledge of the Self, the core essence of our being is available to us all, but where the skills and the motivation to access such knowledge are often suppressed by the acceptance—and often the idealization—of the traditional criteria of success, which emphasize financial gain. Just like the rest of us, leaders have differing abilities to transcend their egos and to gain access to their true, spiritual natures.

Part 5 of the book presents *conclusions* by summarizing the main messages of the contributors of book to stimulate the development of business ethics and sustainability management. One set of messages is related with *research*: how to incorporate spirituality in the theories of business ethics and sustainability management. The other set of messages concentrates on *practice*: how to develop and implement new working models of ethical business and sustainable economy and how to transform leadership to enable ethical and sustainable change.

The book "The Spiritual Dimension of Business Ethics and Sustainability Management" ends with the hope that more and more people will discover how to transcend their materialistic egos, gain access to their true spirituality and find practical solutions to ensure the sustainability of nature and human communities on this Earth.

References

Bouckaert, L., and L. Zsolnai, eds. 2011. *The Palgrave handbook of spirituality and business*. Macmillan: Palgrave.

Bouckaert, L., and L. Zsolnai. 2012. Spirituality and business: An interdisciplinary overview. *Society and Economy* 34 (3): 489–514.

European SPES Forum. 2014: *Definition of spirituality*. http://www.eurospes.be/page.php?LAN=E&FILE=subject&ID=323&PAGE=1. Accessed 27 May 2014.

Fox, W. 1990. *Towards a transpersonal ecology*. Boston: Shambhala Publications.

Heidegger, M. 1978. The question concerning technology. In Martin Heidegger: *Basic writings*, ed. M. Heidegger, 307–342. London: Routledge.

Jonas, H. 1984. *The imperative of responsibility: In search of an ethics for the technological age*. Chicago: University of Chicago Press.

Leopold, A. 1949. *A sand country almanac*. New York: Oxford University Press.

Lovelock, J. 2000. *Gaia: A new look at life on earth*. Oxford: Oxford University Press.

Singer, P. 1995. *Animal liberation*. London: Pimlico.

Soros, G. 1998. *The crisis of global capitalism*. New York: Public Affairs.

Tencati, A., and L. Zsolnai, eds. 2010. *The collaborative enterprise: Creating values for a sustainable world*. Oxford: Peter Lang Academic Publishers.

Young, J. 2002. *Heidegger's later philosophy*. Cambridge: Cambridge University Press.

Zsolnai, L. 2011. Environmental ethics for business sustainability. *International Journal of Social Economics* 38 (11): 892–899.

Laszlo Zsolnai is professor and director of the Business Ethics Center at the Corvinus University of Budapest. He is chairman of the Business Ethics Faculty Group of the CEMS (Community of European Management Schools—The Global Alliance in Management Education). He serves as editor of the "Frontier of Business Ethics" book series at Peter Lang Publishers in Oxford. With Luk Bouckaert he founded the European SPES Forum in Leuven, Belgium. Laszlo Zsolnai's most recent books include "Spirituality as a Public Good" (2007. Antwerp, Belgium, Apeldoorn, the Netherlands: Garant), "Frugality: Rebalancing Material and Spiritual Values in Economic Life" (2008. Oxford, UK: Peter Lang Academic Publishers), "Europe-Asia Dialogue on Business Spirituality" (2008. Antwerp, Belgium, Apeldoorn, the Netherlands: Garant), "Responsible

Decision Making" (2008. New Brunswick, NJ, London, UK: Transaction Publishers), "The Future International Manager: A Vision of the Roles and Duties of Management" (2009. Houndmills, UK, New York, NY: Palgrave Macmillan), and "The Collaborative Enterprise: Creating Values for a Sustainable World" (2010. Oxford, UK: Peter Lang Academic Publishers). His website: http://laszlo-zsolnai.net.

Part II
New Perspectives in Business Ethics and Sustainability

Spirituality: The Missing Link in Business Ethics

Luk Bouckaert

In 2002 I was involved in a discussion with Henk van Luijk, the founder of EBEN, about the future of business ethics.[1] We agreed that at that moment business ethics as a discipline has conquered not only the classrooms of business schools but also the executive suites in many companies. But, simultaneously, we had the feeling that a price has been paid for its success. Business ethics is running the risk of becoming *salonfähig;* losing its critical potential and becoming just part of the business game. I formulated this critical observation in the form of a paradox which says that "the more ethics management, the less ethics in management". The challenging question at the heart of our debate was: How can we overcome this management paradox?

I was convinced that a more spiritual approach to business ethics is needed. Without greater intrinsic motivation, business ethics will sooner or later be reduced to just another instrument for reputation and risk management and any genuine moral commitment will be lost. Henk van Luijk did not agree with my conclusion. He wrote:

> Ethics in business is too serious a matter to make it dependent on the morality and spirituality of individuals (...). Personal moral and spiritual excellence is something we can hope for, but we can hardly influence it. When it comes to active interventions in the domain of business it is the conditions that we should tackle, the institutional configurations that define the range of behavioral alternatives. For business ethics as a discipline and a practice, this entails a substantial broadening and deepening of the field of action. (Bouckaert 2006, p. 204)

I was defending spirituality in business ethics, while Henk van Luijk was pleading for critical reflection on the institutional anchorage of business. In his view the institutional context generates the motivation and not vice versa. Hence, the conclusion that a business-oriented institutional ethics was needed. We must change the supporting socio-political environment if we want more genuine ethics in business

[1] See Zsolnai (2006, pp. 196–222).

L. Bouckaert (✉)
KU Leuven & SPES Academy, Leuven, Belgium
e-mail: Luk.Bouckaert@kuleuven.be

L. Zsolnai (ed.), *The Spiritual Dimension of Business Ethics and Sustainability Management,* DOI 10.1007/978-3-319-11677-8_2

ethics. Of course, spiritual ethics could help with uplifting motivation, but it will not change things substantially.

Looking back at this debate, I am convinced more than ever that business ethics needs a spiritual foundation. The basic reason is that people—due to the current crises—have lost confidence in institutions and institutional leadership. Institutions are part of the problem and not just the remedy for restoring a sustainable future. If we have to reshape our economic, political and religious institutions we need something that has deeper roots than our institutional settings. We need something that can restore a sense of shared meaning, responsibility and purpose. This 'something' is what we may call spirituality. Henk Van Luijk was right in criticizing spirituality as far as it is defined as an introspective and purely individual search for meaning and happiness. This search of individualized spiritual wellness is fine, but will not suffice as a lever for social and institutional change. But we should not reduce spirituality to this private aspects. As a personal and individual experience, spirituality has the power of reconnecting the self from within to all living beings and to the inner Source of Life. Because of this capacity of reconnecting people, spirituality has a strong social and public good character and is linked with the practice of value-driven leadership and with a deep sense of social responsibility.

It would have been fascinating to continue this debate with Henk Van Luijk. In the end we might have reached a meeting point between spirituality and institutional reform. As he died in 2010 I can only present my reflections here as a dedication to his unforgettable commitment to business ethics. In the first section of my paper I will explain the failure of business ethics and the need for spirituality. The other sections link spirituality to leadership (Sect. 2) and to social responsibility and sustainability (Sect. 3). By linking the notion of spirituality to leadership, social responsibility and sustainability, I hope to demonstrate that spirituality is not just an individual matter but a public good that is necessary for leveraging change in institutions.

1 The Failure of Business Ethics

During the 1980s the theory of business ethics was felt to be a revolutionary paradigm shift which could be used as a new horizon, enabling business ethicists to break with the narrowly-focused shareholder theory about firms that are obsessed with maximizing profits. It was the time of the Friedman-Freeman shift, social auditing of The Body Shop, the 'People, Planet and Profit' strategy of Shell, the emergence of ethical investment funds, the growth of CSR programs. It was the time for business ethics to flourish.

The first shocking event that led me to reflect more critically on the assumptions of business ethics was the closing of a Renault plant in Brussels in 1997. Renault at that time had developed an ethos of stakeholder participation and cooperation on the factory floor, but once it was confronted with the problem of profitability, forgot completely about its stakeholder philosophy and fired more than three thousand employees without prior communication or negotiation. This event made me aware of how business leaders—despite their discourse about 'values' and 'stakeholder management'—remained deeply embedded in the logic of maximizing profits for shareholders at the level of strategic decision-making. The failure of Renault was not its search for long term profitability (which is in a free market economy a necessary precondition for flourishing) but the fact that communication and negotiation with primary stakeholders in a context of crisis was not seen as a moral obligation and a consequence of stakeholder philosophy. Why this kind of selective moral blindness? One of the results of this much-discussed case in Belgium was growing distrust of managerial ethical discourse.

The dot-com crisis in the late 1990s (with the fall of big companies such as Enron, Lernaut & Hauspie and WorldCom) highlighted the same paradox: more ethics management does not imply a consistent and integral commitment to ethics. Ethics may be reduced to being applied in a selective and market-driven way. More than the previous scandals, the current financial and debt crisis revealed the limits of business ethics as the (much-praised) practice of moral self-regulation. Although many of the banks involved were also committed to CSR programs and had started ethical investment funds, this did not help them to anticipate and avoid the crisis. Of course, we can explain the ethical deficit as a *lack* of business ethics and see the remedy in more business ethics and more CSR programs. I am afraid that this strategy of 'more of the same' will fail if the efforts made in business ethics are not supported by critical reflection about the mechanisms of selective blindness in business ethics.

Looking at the ongoing financial crisis, it is important to realise that the sub prime-mortgage crisis in the USA was only a trigger. Even without the crisis in the housing market, the system would probably have collapsed sooner or later. Since the 1980s, structural problems have been gnawing away at the global financial system, rendering it very unstable and fragile. The system turned a blind eye to inherent risks and promoted irresponsible, short-term and speculative behaviour. It produced a high degree of *moral myopia* and *selective blindness*. The system not only fostered irresponsible behaviour but also dazzled people so they would not realise the likely consequences or anticipate the looming catastrophe. In classical tragedies the hero's fall is always preceded by his or her inability to grasp the ambiguity of what is happening or the fragility of their predicament. Moral myopia and hubris always come before catastrophe.

If conventional business ethics was not able to help anticipate and avoid the current crisis, will it now be able to develop the right solutions? One of the suggested remedies for overcoming the debt crisis is to launch a new agenda for economic growth following the example of Roosevelt's New Deal after the crisis of the 1930s. It is said that we need a master plan for economic growth to fuel the system, to create employment, consumption and investment, which in the end will enable us to repay our huge deficits and debts. It is not difficult to imagine that such a growth agenda will re-stimulate the overuse of non renewable resources, aggravate climate problems and reduce the quality of life of future generations. Green capitalism may be a step in the right direction, but it is not sufficient. If we build green cars but at the same time stimulate the production and consumption of more cars, we will not stop the overexploitation of our planet. The principle of frugality has to be introduced to help reshape our consumption patterns and life styles. While replacing existing technology with "green technology", we also have to change the incessant underlying drive towards "more and bigger."

If the business ethics paradigm of moral self-regulation through stakeholder-management and CSR programs is not sufficient to overcome the contradictions in our economic system, what can business ethics offer in this context of uncertainty and distrust? We can choose to continue our reformist role within the system as we have done up to now, or we can distance ourselves, apply self-criticism and try to transform our way of looking at things. The latter route was followed by Socrates in Athens, Lao Tzu in Ancient China and The Prophets of Israel. Referring to a more recent example, in his *Guide for the Perplexed* Ernst Schumacher (1977) also did the same at the end of his life. In all these writings we will not find grand theories of leadership and ethics but thoughts about the spiritual way to wisdom, leadership and shared responsibility for the common good. Instead of founding business ethics in the grand rational theories of modernity, such as utilitarianism, Kantianism and social constructivism, we could find inspiration in the older *spirit-driven philosophies* of life and community. They can be very helpful with rediscovering the difference between the *ratio* and the *spirit* as faculties of the human mind. Modern philosophy and education have prioritized human *rationality* at the cost of *spirituality*. Along with many others, I believe that it is time to restore the balance between rationality and spirituality and to re-vitalize our faculty of 'spiritual intelligence' as a source of wisdom in management and leadership.

2 Spiritual-Based Leadership

Instead of defining spiritual-based leadership in an abstract and academic way, let me illustrate it using a quote from the *Tao Te Ching,* where *Lao Tzu* contrasts rational knowledge and leadership with spiritual intelligence and leadership. The *Tao Te Ching* was written in the 6th century BC as a spiritual book for leaders. It contains 81 poems. Here, I quote poem 48 (translated by Stephen Mitchell 1988).

> In pursuit of knowledge,
> every day something is added.
> In the practice of the Tao,
> every day something is dropped.
> Less and less do you need to force things,
> until finally you arrive at non-action.
> When nothing is done,
> nothing is left undone.
>
> True mastery can be gained
> by letting things go their own way.
> It can't be gained by interfering.

According to Lao Tzu, spiritual intelligence is a process of unleashing and letting things go: 'dropping every day something' until you arrive at a stage of *non-action.*

The process is similar to what Socrates has in mind when he writes of the first and most fundamental prerequisite for gaining wisdom: we must arrive at a point of *not knowing*—'I know that I do not know'. While rational knowledge is based on the accumulation of knowledge in order to control our environment, wisdom and spiritual intelligence are based on 'stopping' or transcending rational knowledge in order to open our minds to what Lao Tzu called the Tao, Socrates called the Idea of the Good, Jesus called the Voice of the Spirit and Buddha called the Experience of Emptiness. Whatever we may call this Source of Life and concept of interconnectedness, we will never be able to define or reconstruct it fully using rational precepts or use it in an instrumental way. It can only be referred to as an open Presence in our mind which reconnects us with the flow of Life and to all living beings.[2] This Presence is a source of action but it is not action itself; it is a source of knowledge but not knowledge itself. It is a point in our mind which exists prior to action and rational knowledge that we can identify as a source of energy, inspiration and guidance. Of course, to translate this sense of inspiration into concrete action, practice and institutions we need rational discourse and rational planning. But the challenge is not to crowd out the moment of spiritual intelligence by rationalising our actions and institutions. Spiritual-based leadership is based on the assumption that we can develop spiritual intelligence and allow it to emerge as a non-rational compass for decision making and opening to the future.

In managerial handbooks good leaders are characterized by their ability to define the right goals and to allocate the means available in the most efficient way. Business ethics today is part of the rational profile of the business leader. In order to define 'the right goals' the leader should develop respect for (and conform to) current ethical standards. 'To allocate' means that he or she should incorporate ecological and social costs in the cost benefit calculus of business operations in an efficient and sustainable way. These so-called ethical considerations are requirements for long-term rationality in business. But where do spirituality and spiritual-based leadership come in? Why do we need 'spiritual intelligence' to transcend the rational discourse of business ethics?

I found the philosophical answer to this puzzle in a book written by the French philosopher *Henri Bergson* during the crisis of the 1930s. In his *Les deux sources de la morale et de la religion,* Bergson (1932) identifies two distinct sources of ethics. The first source is social pressure. Societies or collective groups (such as companies) cannot survive if they do not limit the individual opportunism of their members. This is done by means of the development of social taboos, collective rules, internalized moral feelings such as guilt and honor, consultation processes, jurisprudence and so on. Often, religion reinforces these moral rules and feelings by placing them within a perspective of eternity. Following Bergson's view, one can

[2] Because rational and conceptual language always refers to empirical data, or to deductions made from self-evident axioms, we need another, more complex language to express this immaterial and meta-rational experience. We may use indirect, analogical and metaphorical language to refer to the Presence and to the corresponding feelings of openness, co-creativity, love and compassion it awakens.

see business ethics as a form of rationalized group pressure and group control that occurs via ethical codes, moral feelings and consultation processes.

But there is a second source of ethics that Bergson calls "mysticism" which overlaps nicely with what we call "spirituality". Ethical values and rules that help consolidate the collective group may, under different circumstances, become a straitjacket that hinders adaptation to new situations. This gives rise to pressure to change, reinterpret or revise fundamental principles. According to Bergson, such a process of renewal is not a rational one. For in *times* of *crisis* and *transition,* there is no consensus about the basic principles or on the interpretation of these principles. The situation is analogous to what *Thomas Kuhn* (1962) later described in his book *The Structure of Scientific Revolutions* as the prelude to a revolution in science. In such a situation there is no longer a rational basis for justifying the transition from one paradigm to another and this leads to much emotional discussion between groups of scientists, each of whom is convinced that he or she is right.

In turbulent times, fundamental change occurs, in Bergson's view, through moral and religious geniuses/leaders/pioneers who succeed in developing a new way of life in response to a profound *crisis*. People then mobilize around this new way of life and in turn spread and refine it. For Bergson, spiritually-driven leadership plays a key role in the dynamic conception of ethics. From this perspective, he criticizes the classic Kantian and utilitarian theories that claim to have identified fixed norms for good and evil based on universal, unambiguous principles. According to Bergson, these theories overlook the role and example of moral leaders and pioneers. These persons create new interpretations of value and are motivated by three things: first, a keen sense of social frustration and crisis; second, the intuitive and meta-rational sense of the "élan vital" (the inner dynamic) of history; and third, the ability to speak to and mobilize people.

Bergson's account creates a link among ethics, spirituality and leadership. While rational management may suffice in periods of stability and shared trust, it does not suffice in times of deep change. It is no coincidence that nowadays the subject of spirituality in business ethics comes up primarily in the context of crisis, distrust and change. The distinction between the rational manager and the value-driven leader can gain philosophical depth in the light of Bergson's theory of the two sources of ethics.

In current theories of leadership, the distinction between the leader and the manager is mostly linked to the distinction between transformative and transactional leadership (Bass 1990). While the aim of transactional leadership is to motivate and direct people through rewards and punishment, transformational leadership is focused on transforming people by creating a new vision and a shared set of values in an organization. By doing this, the transactional relation is embedded in partnership and open communication which generates trust and intrinsic motivation. Transformational leadership integrates the themes of empowerment, charisma, servant leadership, value-driven leadership or spirit-based leadership.

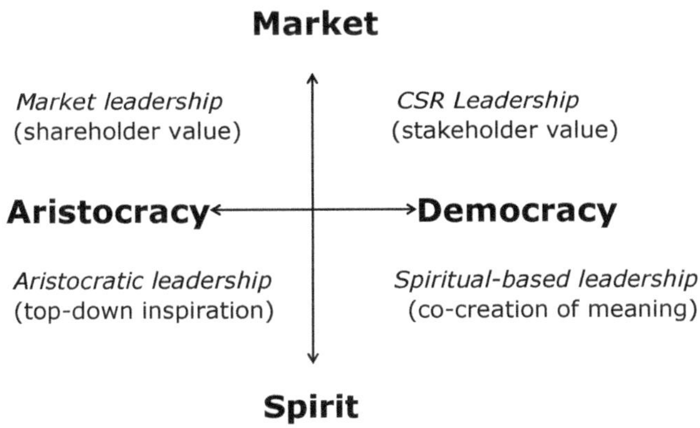

Fig. 1 Types of transformational leadership

However, it is important to realize that spiritual-based leadership is only a qualified form of transformative leadership.[3] Within the scope of transformative leadership there are many types of leaders. Steve Jobs was a charismatic leader gifted with good intuition who was able to motivate people, but his practice of leadership was primarily driven by market imperatives: the creation of shareholder value and strict operational control. Business leaders like Lars Kolind (Oticon) or Muhammed Yunus (Grameen Bank) pay more attention to the co-responsibility and co-creativity of people and to the creation of meaning in work. Another relevant distinction in the field of spiritual-based leadership is the distinction between aristocratic and democratic styles of leadership. Some spiritually-driven leaders follow their religious dogmas and beliefs which often lead to autocratic or paternalistic forms of leadership, while others believe that the Spirit is at work in every person and so support diversity, participation and a bottom up process of decision-making.

The belief that each person has access to spirituality as a source of inspiration and orientation is the core assumption behind spiritual-based leadership. To get more grip on the ambiguities and differences in the field we can think of the structure of the field of transformational leadership as having two axes: a vertical axis representing the tension between market-driven and spirit-centered ideas about leadership and a horizontal axis representing the tension between aristocratic and democratic visions of leadership. Each quadrant refers to a specific type of transformational leadership. Real leaders mostly represent some mix of the ideal types but the typology can help us to realize how different options and styles come together under the umbrella of transformative leadership (Fig. 1).

[3] Books such as *Spirituality and Ethics in Management* (Zsolnai 2004), *U Theory* (Scharmer 2007), *The Soul of a Leader* (Benefiel 2008) and *Leading with Wisdom* (Pruzan and Pruzan Mikkelsen 2007) explore the new paradigm of spiritual-based leadership.

3 Spirituality, Responsibility and Sustainability

Spirituality is often linked to an attitude of being open to the unexpected. This view may lead to the understanding of spirituality as an openness without limits, where 'everything is possible'. However, if we look at the examples of spirit-driven leaders given by Bergson we find that openness of mind is always guided by an *ethic* of *compassion* which implies a deep sensitivity to the vulnerability of life and to people. Creativity without compassion may lead to megalomania or selective blindness about the destructive effects of our behavior. An instructive example of creativity without compassion can be found in Goethe's Faust where in the last act Mephistopheles and Faust develop a visionary colonization project: recovering land from the sea by draining marshlands and building a dike (Bouckaert and Ghesquière 2010). Their megalomania resulted in a kind of blindness about the destructive effects of their project which was exemplified by the death of the elder couple Philemon and Baucis who lived in the area. Faust's behavior can be easily understood as a metaphor for the way we colonize our planet.

Within the framework of rational business ethics, social responsibility is defined as taking into account the rights and interests of *all* stakeholders. With this perspective, being morally responsible means being responsible not only to privileged shareholders but to *all* affected parties. And responsibility is not limited to economic consequences but includes social and ecological consequences as well. As already said, this notion of social responsibility limits and orients entrepreneurial activity in the direction of ecological and social accountability. It is founded in a Kantian concept of universal rights and in the utilitarian principle of maximizing social welfare.[4]

Although there are convincing rational arguments for embedding social responsibility and justice into business, the accumulation of crises reveals that in *practice* we never care about *all* the consequences nor do we take into account *all* the vital interests of stakeholders.[5] In practice we select and interpret rights, interests and utility arguments in order to protect our own interests and to minimize our responsibilities. Our practical sense of responsibility is mostly ego-centric (not necessary egoistical). Many social advocates have called attention to the imbalance between our claims to rights and our commitment to our responsibilities. In order to restore

[4] This theoretical foundation can be illustrated by examining Rawls' theory of justice (Rawls 1971), which demonstrates that under a veil of ignorance a rational individual will promote equality in society and accept inequality only in the case that the situation of the poor can be improved.

[5] Grand rational theories provide a framework for empirical and normative research but simultaneously dazzle and blind us. Rational theories overrate the transparency of reality. How can we identify all the vital interests of all stakeholders? How can we be considered accountable for the consequences of our activities if most of the consequences cannot be known before we act or interact with other people? The problem is that we can claim that grand, rational theories legitimize our activities but actually we are reducing the complexity of reality and the burden of responsibility when we follow our own pragmatic and ideological preferences. Hence there is a deep gap between the claims made by grand theories and real practice. What is missing is a *more realistic and self-critical* sense of responsibility and decision-making.

the balance, they want a charter of universal responsibilities. However, a charter of universal responsibilities runs the risk of remaining a cosmetic operation if we don't succeed in changing our basic attitude towards rights and responsibilities. *Whose rights and which responsibilities must take priority from an ethical point of view*?

From a theoretical and legal point of view everybody has the same rights and the same responsibilities (rights and responsibilities are just two sides of the same coin). However, this is not the case from a genuinely ethical point of view. Emmanuel Levinas (1974) and Hans Jonas (1979)—Jewish philosophers deeply shocked by the Holocaust and the failure of Western ethics to prevent these eruptions of irrational violence—developed after World War II a notion of responsibility that does not start from the point of view of universal rights and principles, nor from a conceptual representation of a global and interdependent world but from the contextual experience of the vulnerability of life.

For Emmanuel Levinas, the original position that awakens our sense of justice and responsibility is not a hypothetical situation under the veil of ignorance, as is the case with Rawls' *Theory of Justice* (Rawls 1971). The original position is a concrete position where I am personally affected by a non chosen confrontation with other peoples' misery and vulnerability. This may happen when I am witness to an accident, or confronted with seeing a demented person losing his or her dignity, or have a conversation with someone who has lost hope. In such traumatic confrontations a feeling of being committed to doing something is awakened and a sense of compassion and responsibility is left behind. Of course I can resist the call to action or the flow of compassion. A powerless and vulnerable person can only touch but not eliminate my freedom and capacity to act. We can neutralize our moral feelings. But the point here is to realize that there is a primary sense of responsibility which is awakened by an immediate and non-conceptual experience of the vulnerability of other people. This contextual experience of vulnerability may grow into a universal ethic of compassion and responsibility.

While according to the philosophy of Levinas the original position is restricted to face-to-face confrontations with other(s), we may enlarge the perspective. Today many people do feel a deep sense of ecological commitment that has been awakened by observing how our planet is fragile and threatened. The effect of this observation of planetary fragility is not only a sentiment of responsibility but a call to act in a responsible way. In this transition from inner feeling to concrete ecological action we need our rationality. We have to conceptualize our intuition, make a tradeoff between different aims and allocate time and scarce means. But what is clear is that there is a spiritual sense of responsibility that precedes the stage of rational conceptualization and implementation. When applying this understanding to business ethics we must not focus too much on the rational foundation underlying the 'CSR principles' but pay more attention to observing the vulnerability of people and the planet. This kind of *sensitive* observation awakens a spiritual commitment to change things.

According to Hans Jonas, the primary sense of responsibility in our age has to be triggered from our modern experience of *fear*. He criticizes the optimistic views of Max Bloch (1959) in his "*Das Prinzip Hoffnung*" that we should imagine and conceptualize the future as a utopian project. In his writings Hans Jonas developed a sense of responsibility generated by a 'heuristic of fear'. Confronted with the planetary impact of modern technology and modern lifestyles we should realize that our planet and the lives of future generations are under threat.

As future generations do not exist as subjects who can claim their rights, they are completely dependent on our good will and are thus extremely vulnerable. We may be concerned for our children, grandchildren and great grandchildren but it is difficult for us to imagine human persons four or five generations away. Hence the more that future generations are distant from us, the more they are voiceless and 'unimaginable'. They have no rights as there is no empirical subject to hold those rights. Nevertheless, as we realize their increasing vulnerability we feel a sense of interconnectedness and responsibility for them. We imagine them as future beings and give them virtual rights. But this is only possible because we have already a notion of responsibility to them, which implies that future generations already exist as objects of our responsibility before they become subjects and holders of rights. In defining sustainability as our responsibility for future generations we should realize that this responsibility is not founded in claims to rights but in the virtual presence of future generations as vulnerable beings.

Sustainability as 'caring for future generations' illustrates very well that, on the one hand, a spiritual commitment anticipates every declaration of rights (and makes such a declaration possible) but, on the other hand, this spiritual commitment must be implemented by giving people rights and by transforming the economy according to these rights. Applied to business ethics this means that stakeholder management and business plans must always be preceded by a spiritual commitment to future generations. I call it a spiritual (and not just a moral) commitment because it does not follow from recognizing existing rights or general principles but rather comes from a personal awareness of the vulnerability of our planet and of future generations. This personal awareness that connects that which is within us to the common good is the first and most intrinsic incentive which can lead us to set up a social praxis of sustainability in business. However, if business leaders fail to take up this challenge, government and the law should protect the virtual rights of future generations and enforce an ethic of sustainability.

4 Conclusion

According to Bergson's theory of leadership the main function of spirituality in business is to open the mind to the *élan vital* of history. This 'élan vital' cannot be seen as a mechanistic or Darwinian program built into the nature of things which may be revealed by positive science. It is instead the infinity of time that creates new meaning in history and in our lives, which we call the Spirit. Spirituality as a faculty of our mind—to be distinguished from rationality—has an intuitive knowledge of this Presence of creativity in life and history. There is always the risk, however, that this intuitive knowledge will be crowded out by a dominance of rational and pragmatic knowledge which is much more focused on problem-solving and controlling our environment.

When we observe the wisdom and practice of spiritual-based leaders in history and modern times, we discover that the openness of their minds is always related to a great sensitivity to the vulnerability of life and future generations. Their spirituality is embedded in a deep sense of social responsibility that drives them to action and into being aware of having a historical mission to accomplish. It is this openness of mind, linked to the idea of a historical mission, which forms the motivational and psychological basis for spiritual-based leadership. At first glance this state of being seems to be the privilege of great historical and prophetic leaders such as Gandhi and Mandela, or in the business world Anita Roddick and Muhammed Yunus. But this is not the case. If we believe that spirituality is accessible for every human being, it can transform every individual mind into a caring attitude for all living beings and generate transformative leadership in day-to-day situations.

Let me illustrate the growing interest in spiritual-based leadership from my own experience of two series of workshops that we organized in the SPES Academy in Leuven, Belgium. The first one was called 'Inspirational Entrepreneurship'. It was designed to explore small and medium size enterprises where spirit-driven entrepreneurship is at work. What inspires these entrepreneurs and how do they translate their inspiration into concrete actions and policies? The other workshop was called 'Thinkers for Doers'. In these sessions we read, together with groups of entrepreneurs, brief, well-selected texts about leadership from great classical authors.[6] I can only witness here that I was surprised that such a significant group of entrepreneurs was eager to develop their spiritual sensitivity and link this knowledge to their business practices.

For those who are afraid that spirituality is just a matter for the happy few, they may be encouraged to remember a quote by the great anthropologist Margaret Mead: "Never doubt that a small group of thoughtful, committed citizens can change the world; indeed, it's the only thing that ever has".

References

Bass, B. M. 1990. From transactional to transformational leadership: Learning to share the vision. *Organizational Dynamics* 1990 (Winter): 19–31.

Benefiel, M. 2008. *The soul of a leader*. New York: The Crossroad Publishing Company.

Bergson, H. 1932. *Les deux sources de la morale et de la religion*. Paris: Presses Universitaires de France.

Bloch, E. 1959. *Das Prinzip Hoffnung*. Frankfurt a. M.: Suhrkamp.

Bouckaert, L. 2006. The ethics management paradox. In *Interdisciplinary yearbook of business ethics,* ed. Laszlo Zsolnai, 199–202, 215–219. Oxford: Peter Lang.

Bouckaert, L., and R. Ghesquière. 2010. Faust and the magic of entrepreneurship. In *Heroes and antiheroes. European literature and the ethics of leadership,* eds. Rita Ghesquière and Knut J. Ims, 13–35. Antwerp: Garant.

[6] For example, texts from authors, such as Lao Tzu, Plato (Socrates), Marcus Aurelius, Ignatius of Loyola, E. Levinas, H. Jonas and A. Einstein.

Bouckaert, L., and L. Zsolnai, eds. 2007. *Spirituality as a public good*. Antwerp: Garant.
Bouckaert, L., and L. Zsolnai, eds. 2011. *The Palgrave handbook of spirituality and business*. London: Palgrave Macmillan.
Jonas, H. 1979. *Das Prinzip Verantwortung: Versuch einer Ethik für die technologische Zivilisation*. Suhrkamp, Frankfurt am Main. (*The Imperative of Responsibility: In Search of an Ethics for the Technological Age*. 1984). Chicago: Chicago University Press.
Kuhn, T. S. 1962. *The structure of scientific revolutions*. Chicago: Chicago University Press.
Lao Tzu. 1988. *Tao the Ching* (trans. by Stephen Mitchell). London: Frances Lincoln.
Levinas, E. 1974. *Totalité et Infini*. Den Haag: Martinus Nijhoff.
Levinas, E. 1996. *Basic Philosophical Writings*. (eds. by A. T. Peperzak, S. Critchley, and R. Bernasconi). Indianapolis: Indiana University Press.
Pruzan, P., and K. Pruzan Mikkelsen. 2007. *Leading with wisdom. Spiritual-based leadership in business*. Sheffield: Greenleaf Publishing.
Rawls, J. 1971. *A theory of justice*. Cambridge: Harvard University Press.
Scharmer, O. C. 2007. *Theory U. leading from the future as it emerges*. Cambridge: MIT Press.
Schumacher, E. F. 1977. *A guide for the perplexed*. London: Harper & Row.
Zsolnai, L., ed. 2004. *Spirituality and ethics in management*. Dordrecht: Kluwer.
Zsolnai, L., ed. 2006. *Interdisciplinary yearbook of business ethics*, 196–222. Oxford: Peter Lang.

Luk Bouckaert is emeritus professor of ethics at the Catholic University of Leuven, Belgium. He is a philosopher and an economist by training. His research and publications fall within the fields of business ethics and spirituality. In 1987 he founded with some colleagues the interdisciplinary Centre for Economics and Ethics at Leuven. In 2000 he started the SPES Forum (Spirituality in Economics and Society) and some years later the European SPES Forum. He wrote several books in Dutch. Recent publications in English include "Spirituality as a Public Good" (co-edited with L. Zsolnai, Antwerp, 2007), "Frugality. Rebalancing Material and Spiritual Values in Economic Life" (co-edited with H. Opdebeeck and L. Zsolnai, Oxford, 2008), "Imagine Europe" (co-edited with J. Eynikel, Antwerp, 2009), "Respect and Economic Democracy" (co-edited with Pasquale Arena, Antwerp, 2010) and "The Palgrave handbook of spirituality and business" (co-edited with L. Zsolnai, London, 2011).

Spiritual Sustainability Management

Hendrik Opdebeeck

Since the first UN Conference on Sustainability in Stockholm (1972), sustainability management (henceforth abbreviated as SM) has had three focal areas: entrepreneurship, the environment and the social needs of present and future generations (Goldsmith 1972). Combining entrepreneurship with respect for socio-environmental priorities is not evident. Therefore not only rationality is necessary but also wisdom (Opdebeeck 2013). Moreover, a recent trend in management that combines sustainability and wisdom is exploring the potentially fruitful bridge between sustainability management and spirituality (Zsolnai 2011). In this paper, the nature of this kind of bridge building is further explored as it appears nowadays to be of central importance to sustainable entrepreneurship that searches for wisdom. During the last few decades Christian and Eastern sources of spirituality like Buddhism and Taoism have frequently proved capable of stimulating the development of SM theories (Opdebeeck 2011). One can also wonder whether and how in the West, Jewish and Islamic sources of wisdom inspire sustainable entrepreneurship.

1 The Jewish and Islamic Tradition

Because of the lack of explicit interest in wisdom of the managerial-minded Romans in the West after Alexander The Great, no new schools of wisdom were created. History speaks of the Dark Ages between Antiquity and the Renaissance. However, during these Dark Ages two great wisdom traditions transferred the rational traditions of Antiquity to the Christian West: Islamic and Jewish culture. The wisdom they transmitted was both spiritual and philosophical. Judaism developed the foundations on which Christian Western culture was later built. Without an understanding of the foundations of the Jewish tradition, Western Culture is incomprehensible.

H. Opdebeeck (✉)
University of Antwerp, Antwerp, Belgium
e-mail: hendrik.opdebeeck@uantwerpen.be

L. Zsolnai (ed.), *The Spiritual Dimension of Business Ethics and Sustainability Management*, DOI 10.1007/978-3-319-11677-8_3

After 70 AD, the Jews emigrated throughout the Middle East, North Africa and Spain and again began to flourish in areas dominated by Islam. Muslims rightly regarded Jewish and Christian precepts to be precursors of their own religion and therefore preached a policy of respect and tolerance for the Jewish and Christian minorities in their midst. Of course, besides this stream of tolerance, both Christians and Muslims also engaged in religious fundamentalism through the Crusades and Jihad. But most historians (including Jewish historians) agree that the glory period of medieval Jewish culture coincided with centuries of tolerance and protection by Islam.

One of the most interesting Jewish philosophers between Antiquity and Renaissance was Moses Maimonides (1135–1206). Among others in his *The Guide for the Perplexed* he transmitted the Greek tradition of practical rational wisdom as the mother of all virtues. Important is the fact that Maimonides and other Jewish philosophers (such as Gabirol), along with Islamic thinkers like Avicenna and Averroes, did more than preserve the rational Greek tradition of wisdom. They also considered how one could bridge rational thought and spiritual traditions in an appropriate and fruitful way. Until the twenty-first Century, as will be explained further, Jewish philosophers like Martin Buber continued with this approach.

2 The Transit of the Greek Rational Tradition to the West

It is remarkable how the above-described transit of Greek rationalism through Jewish and Arabic culture in recent years has been questioned.[1] Authors like Sylvain Gouguenheim (2008) argue that in Europe during the Middle Ages Greek-Latin tradition flourished through translators such as James of Venice, and did not depend on a transit through Jewish and Arabic culture. *In extremis*, the further articulation of the traditions of Greek rationalism during the dark ages could only have been due to the merits of Christian culture. However, even in 1991 the historian Alain de Libera in his *Penser au Moyen Age* resisted this idea of a monopolistic Christian Europe that autonomously inherited the Greek rational tradition (de Libera 1991). It was de Libera who emphatically pointed out that, as indicated above, the West must again become aware of a forgotten legacy. In Toledo in the twelfth century, for instance, translations were being made from Arabic. Scientists at least since the nineteenth century have studied the transit of the Greek rational tradition as embedded in Jewish/Islamic culture.

Worth mentioning in this context is the vision of Pope Benedictus XVI in his Regensburg speech of 2006, *Glaube, Vernunft, Universität* (Verrycken 2011). Ratzinger develops a vision of a complete Greek rational tradition in synthesis with a Bible-based belief in God. One of the most extreme examples of the rationality that Ratzinger supports is his interpretation of the first sentence of the Bible: “In the beginning there was the logos”. God becomes the rationally acting being *par excellence* and is considered to be the rational creator. This Christianity therefore accepts

[1] Inspiration for this chapter was found in K. Verrycken (2011, pp. 963–973).

the analogy between a divine and a human rationality. Furthermore, it is remarkable how Ratzinger insists on the fact that—from this point of view—Christianity may even complete Greek rationality by providing a dialectic bridge between rationality and love, care for others and respect or compassion.

The question then arises, how exactly have other religions that have made a link with Greek rationality come to view this bridge? Besides the bridge Christians have made with the concepts of 'love' or 'compassion', other religions like Judaism and Islam have discovered links with rationality that accentuate other values, such as justice and temperance. Think of the central place of the notion of justice in the Torah (care for widows is an important metaphor) and the Islamic prohibition of the charging of interest as an expression of Muslim temperance. Each year we are remembered of this potential virtue on the occasion of Ramadan. It is interesting to explore whether all these crucial values (compassion, justice, temperance, etc.) could be the very themes we are looking for in our search for a new path for entrepreneurship that combines rational management and spirituality. Hence, to discover the importance of spirituality in SM, it is interesting to investigate how in modern Western spiritual SM, crucial themes such as compassion, justice and temperance are indeed embedded in the concept of rational management.

3 Sustainability Management and the Current Economic Paradigm

It is a remarkable fact that in the SM literature SM primarily questions the basic assumptions of the dominant economic paradigm. These basic assumptions (such as a focus on maximizing the satisfaction of needs and optimizing utility) appear to be grounded on the philosophical foundation of an exclusive belief in rational, quantitative progress. SM therefore elaborates a less one-sided economic paradigm that is aimed at addressing current (and urgent) socio-environmental sustainability problems. The solution of these problems, according to SM, is to be found in the context of calling certain assumptions—like the maximization of the satisfaction of needs and optimizing utility—into question.

Attempts by proponents of the dominant economic paradigm to solve critically important problems on the basis of optimizing utility do not seem to be leading to the eradication of the roots of today's sustainability evils. The sustainable management paradigm even questions whether optimizing utility is valid as a fundamental point of departure. Rather, the case is typically made that today's major socio-environmental sustainability problems may well in essence *be caused* by the central position awarded to assumptions that lead to treating utility as an exclusively important criterion. Hence, the need to elaborate a sustainable economic program that is able to tackle problems using a broader perspective.

Those who seek to elaborate a sustainable economic paradigm to rival the dominant paradigm will also have to develop the core or basic assumptions of that sustainable paradigm in order to solve the socio-environmental crisis in an optimal way. Accordingly, we have to situate SM theories in a wider context and to evaluate existing SM theories by comparing them with other theories that also address how to solve sustainability problems.

4 The Spiritual Dimension Behind Sustainability Management

One specific feature of SM theories is the fact that they often seek a middle way between prioritizing freedom and considering order to be a central concept of sustainable management.[2] Rational management is founded on the concept of developing freedom through the market, but SM rediscovers the need to plan and impose order to limit socio-environmental damage. In other words, SM theories try to guarantee both enough freedom and enough order. On the macro level the choice between either freedom or order risks elevating the tension that exists between those who would promote a market economy and those who prefer a planned economy. This tension at first sight allows no other option than to choose between a laissez-faire system, which avoids planning, and a more totalitarian system, which limits freedom. One could argue that this was perhaps one of the most central problems of Eastern Europe after 1989. Importantly, SM realizes that we are not faced with purely technical problems that can be solved 'rationally'. It thus becomes necessary in entrepreneurship to combine both freedom and order. SM in the first place concentrates on economic practices at the micro-level, realizing that on this level one can, through the discovery and exploration of human self-consciousness, combine freedom and order. Consciousness of the self allows man to generate the (above-mentioned) values such as compassion, justice or temperance, which allow him to transcend the freedom/order dichotomy. When one underlines the importance of this link with virtues such as compassion, justice or temperance, one also understands how SM can indeed be successfully linked with spirituality (Ashar and Lane-Maher 2004; Cavanagh and Bandsuch 2002; Tischler 1999). Moreover, the basis of the typical and often spiritual characteristic of self-consciousness is unique to each individual, and self-conscious individuals may become open to consciously choosing to foster values such as compassion, justice or temperance with a view to making concrete changes in a goal-oriented manner. Such an outcome is possible only if man's inner or spiritual faculties are not neglected. SM, therefore, does not essentially address the structural large-scale macro-economic level, but the personal, small-scale micro-level. Since practice is emphasized above all, one leaves, as it were, the theoretical macro-level to focus on the needed work at the practical micro-level.[3]

---

[2] E.F. Schumacher also worked this out in his "*A*" *Guide for the Perplexed* (1977), probably not coincidentally almost exactly the same title as Maimonides' "The" Guide for the Perplexed.

[3] M. Olson posits that unless the number of persons in a group is relatively small, or unless force or any other means is used to make individuals act in their common interest, rational individuals who have their own interests at heart will not collaborate in the realization of common or group interests (Olson 1994, p. 2).

The emergence of a concrete sustainable economic alternative is therefore linked to the development of practical sustainable experiments that create the possibility for others to be stimulated to turn toward more sustainable practices themselves. This practice can reach the macro level via the meso level. The power of example is vital here. SM attaches the utmost importance to examining experiments with enterprises that exhibit sustainable features. By extension, and after taking any necessary corrections into account, these sustainability features can also be introduced on the meso-level (large-scale enterprises with a more human face, for instance) and on the macro-level (initiatives like Kyoto and Rio+20).

From the perspective of the dominant economic paradigm SM theories are of course deviant theories that point to another economic paradigm. There is not yet plenty of room for a sustainability approach to exist within the dominant paradigm. One can even mention that SM theory often pays too little attention to the political power that plays a part in all the changes that it envisages. No wonder that it is precisely a problem of political will that is the crucial sticking point with platforms like Rio+20. SM risks remaining a vain hope if those who are supposed to put it into practice turn out to have no power, or not enough power. Finally, one can criticize SM even more severely by pointing out that really implementing the sustainable alternative to the dominant economic paradigm would mean the downfall of the existing system.

5 The Utopian Economic Pre-Paradigm

From a historical, economic point of view, SM can be described as rather utopian. Indeed, it is not so difficult to recognize in SM typical elements from the utopian theories of authors like Plato, Saint-Simon, Fourier, Proudhon, Kropotkin, Sismondi, Carlyle, Ruskin, Morris, Hobson, Gandhi and Tawney.[4] These authors are among the great representatives of what (with a nod to Thomas Kuhn) could be called the utopian economic pre-paradigm. When the science of Economics first established itself with Adam Smith's The Wealth of Nations (Smith 1767) at the end of the eighteenth century, alternative paradigms also began to develop. The utopian economic pre-paradigm consists of pre-paradigmatic theories that deal with a more optimal or sustainable economic system that can be imagined and realized. These theories, however, have not yet reached the stage of a fully-fledged "normal" science. Together, they do not, as yet, represent a paradigm of their own. However, they could represent the birth of a new paradigm, so one can call them pre-paradigmatic. Taken together, they seem to represent a utopian economic pre-paradigm.

The term "utopian" means both "as in Utopia" and "of the nature of a utopia". Utopia is the name of an imaginary state in which people live in total happiness, as described by Thomas More in 1516, nearly half a millennium ago. He was referring both to the Greek word "utopia" that literally means "no place", and to another

[4] For a detailed survey, see F. Manuel (1979).

Greek word "eutopia" that means "a good place". A utopia is, however, often wrongly understood to be a dream of a perfect state which will never be realized. When SM theories are described as utopian, "utopian" is interpreted according to the meaning just described. SM is then thought of as a naïve movement that devotes all her energy to creating an imaginary system that perhaps never will be realized in actual practice.

Yet SM situated in what we call the utopian economic pre-paradigm is something that can be realized, and modern reality proves this claim. It represents a more optimal kind of management that does not fully exist as yet, as in the case of Moore's Utopia. The utopian pre-paradigm finds its foundation in the many utopian thought-constructs that have been elaborated since the formerly-mentioned Greek rational tradition. Think of Plato's writings and what happened in Sparta. Later on, important theories within this pre-paradigm were constructed in the context of utopian socialism (Saint-Simon, Fourier, and Owen), anarchism (Proudhon and Kropotkin), and humanistic socialism (Sismondi, Ruskin, and Hobson). Typical of all these utopian theories is their goal of combining rational practice and the aforementioned values such as compassion, justice and temperance, in the West transmitted or strengthened by religious traditions like Christianity, Judaism and Islam.

6 Martin Buber on Utopian Theories Versus Marxist Theories

Before we compare SM with some important currents within the utopian economic pre-paradigm, we would like to insist that an essential difference exists between SM and Marxism. Ever since Marx and Engels described thinkers like Saint-Simon, Fourier, and Owen as "utopian socialists" in the 1848 Communist Manifesto, the word "utopian" has had rather negative connotations. Later on, Marx and Engels were to resist the typically utopian approach of thinkers such as Proudhon, a representative of anarchism. Marxism accuses the utopian pre-paradigm of relying too much on a combination of rationality and a sense of compassion, justice and temperance that belongs to the realm of self-consciousness or spirituality. Marxism, on the other hand, lets history do its work and states what is dialectically happening in history as far as the conditions for production are concerned.

Where Marxism believes that these dialectics produce revolutionary class struggle, the utopian paradigm believes (as explained by the twentieth century Jewish philosopher Buber), that if one strives for something now, one has to create the possible space for it now, so that it can be implemented then (Buber 1950). Marxism believes this approach is doomed to failure. For Marxism, any attempt to change society gradually (e.g. starting from the micro level, like SM does) amounts to a refusal to acknowledge how history evolves. Marxists believe that history can be described in a purely objective and rational way without reference to ethical or spiritual elements. This is why Marxism, which understood itself as "scientific socialism", distanced itself from "utopian socialism", and later also from other utopian currents. The rather more utopian (not to say, sustainable) passages in Marx' writings, such as the diatribe against alienation and the emphasis given to the importance of decentralization, eventually disappear into the shadow cast by his scientific socialism in which centralized revolution appears as the leading concept. In the end, as is well-known, scientific socialism was to get the better of utopian socialism. The Marxist paradigm was realized in practice (for instance, in Eastern Europe), whereas the utopian paradigm remains as yet on the drawing board. Both paradigms are complex and geared to problem solving, but they start from different points of departure. Marxism starts at the macro level, utopian socialism at the micro level.

7 Sustainability Management as an Utopian Pre-Paradigm

A lot of characteristics of the utopian economic pre-paradigm can also be found in SM theories, albeit with a specific emphasis all of their own.

Searching for a more optimal role for government, for instance, as a promoter of sustainability, has always been of vital importance in SM theories, as it was in the earliest utopian theories that started with Sparta in Greek antiquity. Sparta's condemnation of a rich, luxurious state can be seen as an extreme form of what SM nowadays understands as the economics of frugality. As a matter of fact, following Plato (who established principles designed to lead individuals to create a better society), SM also often emphasizes that the starting point for change is the personal micro-level.[5]

In his Utopia, Thomas More radically rejects the notion of private property. SM also deals with the question of what the appropriate type of property is. Think of the importance of cooperatives like Mondragon as instruments of sustainability. We find similar statements in the works of the eighteenth century forerunners of utopian socialism, Morelly and Godwin. Godwin asked the question whether man's instinctive desire for property could, in practice, be limited by sound education. SM also considers education and teaching necessary but not sufficient preconditions for changing human mentality and promoting sustainable behavior.[6] Teaching and education should be supplemented by embedding them in a stimulating context (e.g. the state, as mentioned above) that testifies, in a concrete way, how things can be done differently and how these differences can be brought to bear on the macro level.

One of the central themes in the utopian work Francis Bacon developed to improve social systems concerns the best possible application of science with a view to optimizing technology (or, more concretely, techniques for production). SM also emphasizes the importance of the best possible techniques of production, but it no longer possesses Bacon's unlimited faith in the progress of science. For SM,

[5] The most important point for Plato was that man should live according to those principles, but Plato himself was not at all sure of the practical feasibility of the utopian state he elaborated. See Plato (1973, pp. 368–369).

[6] Saint-Simon opted to provide free education to the children of workers to stimulate forms of worker co-ownership. In the context of cooperatives, Owen also emphasized the importance of providing education to the children of workers.

modern problems require more appropriate forms of technology. Later, Kropotkin also advocated suitable types of technology. Morris, and another one of Ruskin's disciples, Gandhi (1869–1948), had very clear thoughts about this subject, both for industrialized and less-developed economies.[7] It was also Gandhi who worked out a profound and concrete vision of how to fundamentally respect the environment, a basic priority we have mentioned as a pre-requisite for the practice of SM.

An important concept that also links SM to utopian socialism is the revaluation of labor. Charles Fourier (1772–1837) and Robert Owen (1771–1859) wanted to establish a real alternative to traditional labor relationships on the micro level by means of their production cooperatives. Similarly, SM emphasizes that labor should not only result in the production of goods and services that are required, but also in the development of creativity and the structure of cooperation with others. The type of self-management combined with division of profit advocated by Louis Blanc (1811–1882) in his production cooperatives is very similar to SM's ideas about the topic.

In extremis, SM could be linked to anarchism by the emphasis SM puts on the necessity of reducing hierarchical society by stimulating the creation of a society that is built from the bottom up. Instead of simply being part of an over-centralized society, people could, on the micro-level, form groups on a voluntary basis, as advocated by Proudhon (1809–1865). Martin Buber uses the concept of a-cratism (the absence of dominion), rather than anarchism (the absence of rule) in this context (Buber 1950).[8] Some components of the concept of self-ownership were elaborated in Proudhon's basic thoughts about cooperatives in a manner not dissimilar to those pursued by SM, which promotes the evolution of democratic co-management (Proudhon 1926).

At first sight, like anarchism, SM also implies resistance against various forms of large-scale enterprises.[9] As with Kropotkin (1842–1921), SM remains opposed to all types of mass production resulting from the increased application of large-scale technologies. SM also often rejects the far-reaching division between manual and intellectual labor that is the result of large-scale production.[10] However, SM fundamentally differs from anarchism in that SM is always interested in trying to promote sustainability in an economic system more geared to man and the environment in large-scale enterprises as well. For SM, an exclusive focus on

---

[7] After Gandhi came to know Ruskin's work as a student in England he described Ruskin's Unto This Last as "the one book that brought about an instantaneous practical transformation in my life" (Gandhi 1927, p. 22). Hobson further developed Ruskin's insights for the industrialized world. Gandhi did this for the less-developed world. Gandhi stated, on the subject of modern technology: "I am not fighting machinery as such, but the madness of thinking that machinery saves labor. Men save labor although thousands of them are without work and die of hunger on the streets(…). At the present the machine is helping a small minority to live on the exploitation of the masses" (quoted in Madan 1927, p. 134).

[8] In another strand of the utopian economic pre-paradigm, humanistic socialism, William Morris clearly posited the need for small, clearly delineated units of administration.

[9] J. de Sismondi (1773–1842), another adherent of humanistic socialism, was vehemently opposed to any form of excessively large-scale industrialization.

[10] In the tradition of anarchism, L. Kohr states in his book, (1957), that the most important economic problems of our time are caused by the excessive sizes of nations.

small-scale enterprises represents another extreme.[11] SM always has to emphasize and repeat that her theories do not just call for small-scale enterprises. SM emphasizes the importance of appropriate contexts that incite individuals to collaborate with others within an economic system in which man and environment occupy a central position.

8 Conclusion

Utopia and sustainability management are concerned with allowing various human potentialities to come to fruition in a more optimal environment and society. They are creating the possible space in the 'now' for what they want to achieve in the 'then'. The fundamental difference between the approach taken by spiritual oriented SM and the utopian economic pre-paradigm can be expressed as the difference between eschatology and utopia.[12] Eschatology is concerned with the final completion of creation, or even of evolution. The Jewish thinker Martin Buber describes the difference as follows:

> Even though eschatology, especially in its most basic, prophetic form, awards man an important, active share (…), the decisive act takes place above man. In utopian thinking, on the other hand, everything is subject to man's conscious will. One might even describe utopia as an imaginary image of society constructed as if there were no other elements involved except for man's conscious will. (Buber 1950, p. 17)

However, it is entirely understandable that not everybody is at ease with a spiritual background of SM. The impact of SM in society might be limited by this emphasis. This is because one might be tempted to think that realizing sustainability in management cannot occur without a spiritual background (Holthaus 2008). Yet a sustainable system, as proposed by SM, may appeal to both those who are interested in spirituality and those who are not (or are less so). Consequently, the formerly-described strong link between SM and the utopian gift is of the greatest importance. History has proven that the utopian pre-paradigm offers a possibility to elaborate crucial aspects of sustainability within a pluralistic context, both explicitly spiritual and a-spiritual. In fact, in the described utopian paradigm the essential values required to make sustainability concrete (like compassion, justice and temperance) are inspired by articulated philosophical wisdom from Greek Antiquity and by spiritual traditions like Christianity, Judaism and Islam, not to forget Eastern traditions like Buddhism or Taoism. From a historical utopian background it becomes clear how sustainable management as such *can* be the current utopian denominator by which management is made more in balance with the environment and man. For this kind of management, wisdom is necessary and spirituality can be more than helpful.

---

[11] It should be admitted though that Kropotkin also recognized *de facto* that large scale operations are indispensable for undertaking a number of activities.

[12] Even if one were to try to locate authors belonging to the Christian solidarity school (like J. Buchez (1796–1865), F. Huet (1814–1869), and Charles Kingsley (1819–1875)) inside the utopian economic pre-paradigm, this Christian source of inspiration remains a marginal influence in the greater whole of pre-paradigmatic utopian theories.

References

Ashar, H., and M. Lane-Maher. 2004. Success and spirituality in the new business paradigm. *Journal of Management Inquiry* 13 (3): 249–260.
Buber, M. 1950. *Pfade in Utopia*. Heidelberg: Schneider.
Cavanagh, G., and M. Bandsuch. 2002. Virtue as a benchmark for spirituality in business. *Journal of Business Ethics* 38:109–117.
de Libera, A. 1991. *Penser au Moyen Age*. Paris: Seuil.
Gandhi, M. 1927. *An autobiography*. Ahinedabad: Navajivan.
Goldsmith, E. 1972. A blueprint for survival. *The Ecologist* 2 (1, Special Issue): 35.
Gouguenheim, S. 2008. *Aristote au Mont-Saint-Michel. Les racines grecques de l'Europe chrétienne*. Paris: Seuil.
Holthaus, G. 2008. *Learning native wisdom. What traditional cultures teach us about subsistence, sustainability and spirituality*. Lexington: University Press of Kentucky.
Kohr, L. 1957. *The breakdown of nations*. New York: Rinehart.
Madan, G. 1966. *Economic thinking in India,* 134. Delhi: Chand.
Manuel, F. 1979. *Utopian thought in the western world*. Oxford: Blackwell.
Olson, M. 1974. *The logic of collective action*. Cambridge: Harvard University Press.
Opdebeeck, H. 2011. Compassion: Chinese and western perspectives on practical wisdom in management. *Journal of Management Development* 30 (7–8): 778–788.
Opdebeeck, H., ed. 2013. *Responsible economics, E.F. Schumacher and His legacy for the 21st Century*. Oxford: Peter Lang.
Plato. 1973. *The republic*. Harmondsworth: Penguin Books.
Proudhon, P. 1926. *"Confessions d'un révolutionnaire" Oeuvres complètes*. (eds. M. Bouglé and H. Moysset). Paris: Société des amis de Proudhon.
Schumacher, E. 1977. *A guide for the perplexed*. London: Jonathan Cape.
Smith, A. 1981(1767). *An inquiry into the nature and causes of the wealth of nations*. Indianapolis: Liberty Fund.
Tischler, L. 1999. The growing interest in spirituality in business. A long-term socio-economic explanation. *Journal of Organizational Change Management* 12 (4): 273–279.
Verrycken, K. 2011. Aristoteles op de Mont-Saint-Michel, Reflecties bij de Gouguenheimaffaire. *Streven* 14 (2011 December): 963–973.
Zsolnai, L., ed. 2011. *Spirituality and ethics in management*. Dordrecht: Springer.

Hendrik Opdebeeck is professor of Philosophy and Economics at the University of Antwerp where he is affiliated to the Centre of Ethics. He studied philosophy and economics at the universities of Leuven and Gent where he obtained a PhD with a dissertation on E.F. Schumacher (1911–1977) and the polyparadigmatic discourse in economics. His research interest is focused on the cultural-philosophical backgrounds and effects of globalisation with special attention to the role of spirituality and technology. In 2000 together with Luc Bouckaert he founded the SPES Forum for Spirituality in Economics and Society in Leuven, Belgium.

The Inner Perspective—The Sufi Approach

András László

The paper explores the inner perspective of leadership and sustainability so seldom touched upon. First, a concise introduction to Sufism and the teachings of *Sufi Masters Hazrat* and *Pir Vilayat Inayat Khan* will set the tone and framework within which the focus on the missing link—*the Inner Perspective*—in Sustainability & Leadership literature will gradually unfold.

The introduction will highlight:

- new settings/dimensions of consciousness according to Sufism
- the transcendent dimension: awakening from personal consciousness
- the spiritual dimension of emotions
- the essence of creativity: tapping in to the thinking of the Universe
- how the shift of perspective—turning within—results in overall transformation

Within this outlined Sufi framework, the paper will continue to explore the idea of the much-needed *leadershift* in our age of mass collaboration and ever-deepening interconnectivity. It points to the *ethics of sustainability* as well, and, as regards *spiritual leadership* and *creativity*, it will put in the forefront the recovery of the soul, the heart and wisdom as pivotal added value. It concludes by highlighting and pinpointing the essence of spiritual leadership/the leader shift: namely, the identification of "that what transpires behind that which appears".

1 What is Sufism?

Sufism is not easily defined, for it is neither a distinct religion, or cult, nor a distinct and definite doctrine. It does not require adherence to prescribed dogma or ritual and it welcomes members of all races, nations and beliefs. It has existed throughout

A. László (✉)
GlobalVisioning.net, Brussels, Belgium
e-mail: alaszlo@telenet.be

L. Zsolnai (ed.), *The Spiritual Dimension of Business Ethics and Sustainability Management*, DOI 10.1007/978-3-319-11677-8_4

the centuries in many cultures. There is no single founder of Sufism but it has been incorporated through the ages into teachings originating from Zoroastrianism, Buddhism, ancient Egypt, Greece, and India around its Islamic kernel.

Sufism is a *way of looking at the world* and a *way of living in the world.* It emphasizes bringing one's highest ideals into everyday practice. It is a way for humanity to awaken to the abundance of life, both individually and collectively in every area of human enterprise.

Great Sufis have appeared at different times and have founded schools of thought. Their expression of wisdom has differed to suit their environments, but their understanding of life has been one and the same. Every age of the world has seen awakened souls, and as it is impossible to limit wisdom to any one period or place, so it is impossible to date the origin of Sufism.

The main ideal of the Sufi school is to attain that perfection which Jesus Christ taught in the Bible, "Be ye perfect as your Father in Heaven is perfect". The method of the Sufis has always been that of *self-effacement*; but the effacement of which self? Not that of the real self but of the false self (on which man depends, priding him or herself on being something in order to allow that real self to manifest in the world of appearances). The Sufi method works toward the unfolding of the soul: the self which is eternal, to which all power and beauty belong.

Sufism was intellectually born in Arabia, devotionally reared in Persia, and spiritually completed in India.

The term "Sufi" may be derived etymologically from the Greek *Sophia*, wisdom, which is acquired both from inside and outside. Some, however, claim that its origins lie in the word *suf*, the woolen garments worn by the *fukran*, who are those that observe austerity.

The Founder of the Sufi Order in the West—Hazrat Inayat Khan

The Sufi Order evolves from the universal spiritual tradition of the Chishti lineage which originated in the East and was brought to the West by *Hazrat Inayat Khan*, the founder of the Sufi Order in the West, in London in 1916. He was given the task by his Sufi teacher to "unite East with West in the harmony of your music" (*Hazrat Inayat Khan*—a renowned singer/vina player—was awarded the greatest musical title in India: Tansen of India). As the first Sufi teacher in the West, Hazrat Inayat Khan sought to make the spiritual legacy of Sufism responsive to the needs of our time. He landed in America and later traveled to Europe and Russia, sowing the seeds of Sufism. He was one of the first to speak of an emerging planetary consciousness as the next stage in the spiritual evolution of humanity. In the later years of his life he spoke of Sufism as a mother who would give birth to child whom he called 'the Message'; that is, beyond name and label. He believed the Message would facilitate the *awakening* of the *consciousness* of *humanity* to the *divinity within*, and bring new life to all facets of human endeavor.

The Spiritual Successor and Head of the Sufi Order in the West—Pir Vilayat Inayat Khan

Pir Vilayat Inayat Khan, son of Hazrat Inayat Khan, was born in London in 1916. His early years were imbued with both the rich mystical tradition of the East and the heritage of the West. His later training also reflects the synthesis of East and West. He studied philosophy and graduated with a degree in psychology from Paris University; he later did postgraduate work at Oxford and also studied music at l'Ecole Normale de Musique de Paris. He then began intensive meditation practice in India and the Middle East with Sufi masters and teachers of various other meditative disciplines, and undertook long periods of seclusion and retreat. In 1926, his father named him his successor and head of the Sufi Order and the Confraternity of the Message. In accordance with his background, Pir Vilayat exemplifies the global consciousness of the emerging holistic age, allowing him to fulfill his purpose and present the message of unity. All his life, Pir Vilayat has been well-known as a master of meditation who has integrated the ancient traditions of meditation practice with the relevant creative forces at work in today's world. He has evolved a holistic approach to uplifting and expanding consciousness and then incorporating heightened awareness and insight into everyday life situations to find meaningfulness in activity and an understanding of one's purpose in life.

2 The Awakening of Humanity to the Divinity of Man

> The solution to the problem of the day is the awakening of the consciousness of humanity to the divinity of man. (Khan 1978, p. 142)

The human soul is continually seeking to realize its true being and to fulfill its purpose in life. This can be seen not only in each individual's yearning to realize to the fullest the potentialities inherent in his or her being, but also in our collective being's unfolding towards the highest expression of human existence. The consciousness of humanity is awakening to the interrelatedness of all of life, and this is reflected in the new holistic understanding of the universe that is emerging from all spheres of human experience today. We can see this trend in the growing concern for the ecological balance of our blue planet, interest in holistic health, a deeper social conscience that is concerned with the plight of suffering people all over the earth and a growing awareness of the one underlying truth which is promoted through all the religious traditions. At the same time, recent scientific discoveries are disrupting established concepts and models of the universe. In physics, biology, medicine, and psychology the very foundations of our notions about life are being superseded by the realization that we are both fractions of the universe linked with all the other fractions by their resonance, but also the totality as every cell of our body contains the code of the whole universe. This view, in fact, corroborates the vision of the mystics throughout history who have peered beyond the frontiers of their separate and limited personal perspectives to encompass the vast reality of the whole.

This is the way of the Sufi, which views life in the light of the one essence behind the surface of life; the essence revealed by all the religions.

Pir Vilayat Inayat Khan highlights the words of wisdom of his father by explaining that

> …what Hazrat Inayat Khan calls the awakening of the soul of humanity to the consciousness of the divinity of man is something that is happening on a large scale and we all are part of it. Consequently, our meditation can no longer be confined within one person, even to liberate ourselves from one person, because together we constitute a reality that is more than the sum of its parts. This is God being born in our midst, in the convergence of our souls. The most meaningful thing in life is that something is gained by existence, by the interfacing of us. The heavens are enriched by the experience of the earth; in other words, God is enriched by His experience of becoming man. Thus, the new orientation of the Sufis is emphasized in Hazrat Inayat Khan's teaching on the art of personality and God's fulfillment in existence in and through man. The purpose of man is like the horizon: the farther he advances, the farther it recedes. There comes a time when one discovers a cosmic purpose beyond one's personal purpose. This is exactly the realization that was outlined by Hazrat Inayat Khan as the motto of his age: the awakening of humanity to the divinity of man. (Khan 1982, p. 48)

Following this concise introduction to Sufism and the teachings of Sufi masters Hazrat Inayat Khan and Pir Vilayat Inayat Khan, allow me to present—as a kind of transition—four wonderful and insightful quotes:

> Thinking is the enemy of creativity. Ray Bradbury (1953, p. 55)

> The problems of the world cannot possibly be solved by skeptics and cynics…We need people who can dream things that never were. John F. Kennedy (1964, p. 137)

> The most important thing going forward is to break the boundaries between people so we can operate as a single intelligence. David Bohm (1985, p. 44)

> The dream of one person is just a dream. Dreaming together is the start of a new reality. Friedensreich Hundertwasser (2004, p. 78)

3 Creativity

"Every human being creates as he/she breathes", André Malraux once said. A drawing of a child, a sonata from Mozart, a new car… they all are hallmarked by creativity. But what does it mean to be creative? Is creativity a gift, a genetic or cultural endowment, or the result of environmental conditioning?

For societies (as well as for each one of us) to be creative means first of all having an ardent desire to live. Creating in everyday life means inventing a world in which one feels happy to live, together with others.

To create means "to bring into existence". Creativity is essentially an internal, ongoing process within each of us. As *Carl Rogers* pointed out so accurately in his book, *On Becoming a Person*, "The creative process is the emergence in action of a novel relational product, growing out of the uniqueness of the individual on the one hand, and the materials, events, people, or circumstances of this life on the other.

The mainspring of creativity appears to be man's tendency to actualize himself, to become his potentialities" (Rogers 1977, p. 27).

Shift in Perception

In the 3rd millennium, the necessity for the transformation of our goals, perceptions, feelings, attitudes, behaviors, communication patterns and values behind our judgments and decisions is intensifying and is of pivotal importance.

Today's economic, social, cultural and ecological challenges can only be met powerfully and effectively through a *shift* in *consciousness* and *awareness* which in fact is already well under way today. The good news is that more and more people today are becoming aware that we have fundamental challenges to overcome—in our families, our emotional systems, our cities, our governments, our economic systems, our international relations and our planetary environment—and that we must find new ways to embrace these problems.

Marcel Proust once said, "The real act of discovery consists not in finding new lands but in seeing with new eyes".

In his seminal book *The forgotten half of change—achieving greater creativity through changes in perception*, author *Luc de Brabandere,* Europe's best-selling author on business innovation and a partner in the Boston Consulting Group, explains the difference between innovation and creativity (Brabandere 2005). Changing reality is *innovation.* This requires action by a team, is continuous, takes a long time, and delivers something new to the system. Its impact is measurable and certain and it requires the use of tools and project management. *Creativity,* on the other hand, changes perception and requires thinking/feeling. It is a challenge for individuals and it is discontinuous, taking an instant. Its impact cannot be measured, and it requires learning methods fuelled by questions, surprises, and incomplete answers.

Luc de Brabandere further argues that "to really change, we have to change twice. Not only do we have to change things, but we have to change the way we see things. Innovation, thus, is people's capacity to change reality, and creativity is the capacity of people to change their perception of reality" (Brabandere 2005, p. xi).

Change in *perception*—the instant that a new idea occurs, creative and disruptive. The key to this change in perspective and to inner leadership is *turning within.*

Turning Within

Millions of people the world over are trying to transform and improve their lives. More and more are becoming aware that the answer to changing the world does not come from a source outside of them. The answer comes from within: to discover the whole world outside of us as part of us, to awaken to the myriad of potentialities latent in us that are waiting to be activated and come through in our personalities.

In confronting oneself, taking responsibility for one's own thoughts, feelings, emotions and actions, one touches upon a level of awareness where one experiences unity with all the living through love and compassion and the power of creativity.

Turning within thus means a shift in attunement, in perspective. Withdrawing your awareness from the physical and mental planes and focusing it on the inner depth of your being. As you go deep within yourself you discover a whole new mode, a whole new dimension of thinking/feeling which you never encompass in your ordinary way of thinking. You also discover dimensions of your being that are entirely different from the ones with which you normally identify. To look inside yourself means reaching beyond the limits and limitations of your own self-image. It is an internal way of looking at things, thoughts, emotions, opinions, people, problems and situations. By practicing it continuously in everyday life, one has the chance to see connections between things which did not seem to be connected. One can see the interrelatedness of every person and thing, the interdependence rather than the independence, and also the separateness of events/individuals—which is what you normally see when you are considering the world as it appears from outside.

The inside is governed by resonance, affinity, and a very fine attunement: everything is interspersed with everything else. Everything is delicately interwoven, like a net, an *inner-net*.

As a good illustration of this interplay of outside-inside you could consider the difference between swimming at the surface of a lake and seeing the separate lotus flowers, or swimming under the surface and seeing that the flowers are interconnected, all part of a large network. In the deeper, more impersonal mode of thinking you experience receptiveness, openness to all the possibilities and qualities that are waiting and wanting to come through and enrich our personalities. Reaching out from inside removes the obstacles standing in the way of us unfolding the potentialities that are lying in wait in our being, and touches upon more vast and subtle levels of our being.

Inner Leadership

The self-organizing creativity emerging from within and interfacing with the environment makes all the difference. This is inner leadership.

In this mode of thinking and awareness experience has become self-discovery. It means exploring through your creative imagination what you would have been if you could have been what you might have been.

Creativity is our birthright. We all are born with the unlimited potential to be creative, to display this potential, and by doing so contribute to the development of the unlimited potential of our universe in ongoing transformation.

With each creative act one has the impression that suddenly one's horizon has been broadened. Limits have disappeared; fears have been transmuted into new possibilities, another frame of reference has emerged which goes with an increasing sense of autonomy, confidence and marvel. This is inner leadership.

To create and to be an inner leader means to calling upon the best in oneself, with love, compassion, sensitivity, imagination and inspiration. Constantly being alert and open to that which presents itself, transpires through that which appears and wants to emerge. The creative being and leader is always ready to transgress, to go beyond that which confines in order to awaken that which lifts up. Obstacles are viewed as challenges, errors as precious information, setbacks as opportunities to break through.

Yet, more than ever, and generation after generation, increasing numbers of people find themselves willy-nilly confined by the inexorable machinery that conditions them and robs them of their creativity and (inner) leadership.

At what cost our modern efficiency?!

People suffer from a total lack of opportunity to harness their creative urges, unlike the time before arts and crafts were replaced by (machine-driven) standardization—now computerized. Our experiments with political institutions have proven how crucial freedom is to the promotion of progress. In fact, freedom from conformity generates creativity and leadership. On the other hand, there is nothing in the world more abused than creativity. We see the consequence of this in the decadence, vulgarity, carelessness and permissiveness of our modern societies.

There is no accounting for taste!

To quote Pir Vilayat Inayat Khan:

> One could define creativity as the act of exploring unchartered regions of the mind while grasping a correspondence between the mental constructs thus gleaned and a form of configurations or scenarios in the fabric of matter. Creativity is then a congruent conjunction between the timeless and the transient, the heavenly and the earthly. It is a sudden sense of meaningfulness that sparks our innovative faculty. What is meant by our sense of meaningfulness is our mind's ability to click when it grasps a correspondence between two thoughts which had hitherto appeared unrelated. To be creative rather than just fanciful, one's vision needs to click with the possibility of its actual realization at the existential level. (Khan 1982, p. 51)

Creativity and Inner Leadership in Everyday Life

How can one create in everyday life that radiant ability for renewal, that liberty called creativity? Does it not, first and foremost, involve a revolution of one's perceptions and attunement?

To see and to resonate without purpose, without preconceived ideas, without desires and feelings distorting the reflection in the mirror of the eye.

To really see; that is, to unveil.

How can we create the conditions necessary for creativity and leadership to emerge in inner openness and silence?

The answer has everything to do with broadening, uplifting our consciousness, focusing it on the way we live, behave and act in everyday life. First of all, it requires alert attention to every instant so that deep creativity can burst forth as freedom, enabling us to welcome any change with a complete inner receptiveness. To create and be a leader, then, means being able to meet any situation in an original and fresh way. And the spectacular thing about such a moment is discovering the boundless receptiveness in the glance of another person.

Creativity (and leadership) is not about reacting to the environment, or processing it, nor has creativity been promoted or even catalyzed by anything from outside. We do not know what the system of the process of association is that led to it; it may have been catalyzed by a situation, a thought or a feeling, but it is not a reaction.

One of Europe's most sought after leadership speakers, *Emmanuel Gobillot*, explores in his latest groundbreaking and fast-paced book *Leadershift—Reinventing Leadership for the Age of Mass Collaboration* the world of mass collaboration (meaning: the collective actions of large number of people working independently of organizations) and what it means for leaders (Gobillot 2011). He twice quotes French author Antoine de Saint-Exupéry (1984). The first quote goes as follows: "As for the future, your task is not to foresee it, but to enable it". According to Saint-Exupéry, most of what leaders (still) do is engineering. Faced with any narrative, the usual leadership response still remains the same: find, attract, nurture and convince by engineering (or that other fad, reengineering) solutions to problems. Mass collaboration requires, precisely, that we find a new *modus operandi*. Gobillot continues his argumentation as follows:

> Our leadership models have been influenced by our understanding of the psychology of both leaders and followers. We look for deficits that need to be fixed rather than strengths that can be augmented. We focus on why individuals do what they do, hoping that by deepening our understanding we can engineer better control mechanisms. Leading in mass participation requires us to become social anthropologists, focusing on the communities that are created, rather than the psychology of any one individual who contributes to one or more of them. The aim of leadership will always be to secure engagement, alignment, accountability and commitment. What is changing is the way in which we will achieve these. If you ever feel the world of leadership is getting tougher it is because it is. The reason this is being felt like never before is that the world opening in front of us is making many of our tried and tested success recipes redundant. (Gobillot 2011, pp. 5–6)

Gobillot's closing reflections start by quoting Antoine de Saint-Exupéry again from *Wind, Sand and Stars*: "What saves a man is to take a step. Then another step. It is always the same step, but you have to take it". (Saint-Exupéry 1984, p. 23) Gobillot then argues that…

> I can't help but think that the ultimate leadership challenge is not the erosion of the powers and tools thrust upon us by a turbulent environment. The ultimate challenge for any of us is the ability to take Saint-Exupéry's first step. We cannot second-guess the future. There is no point looking for a truth that will answer all our concerns. It is not out there. Our job as leaders is to take the first step, without trying to second-guess or fearing what might lie ahead. Our future lies in our ability to march proudly into our future—at some times leaders and at others followers—working together, building on each others' strengths. (Gobillot 2011, p. 160)

Creativity and inner leadership is—in essence—the capability of the human being to tap in to the thinking of the universe which is the greatest miracle. The exploratory drive in human beings is crucial because this is, precisely, creativity. It is trying out new ways of doing things and new combinations.

Meaningfulness

Meaningfulness is the true criterion of creativity and inner leadership.

It is meaningfulness that makes us choose between fantasy and imagination. The question of meaningfulness is very important in intuition, and when we are creating it is our sense of authenticity that is at work.

What makes somebody creative? The incredible amount of energy, the flexibility, the curiosity, the sensitivity as well as the willingness to constantly challenge and question him/herself. Creativity brings the whole person into play. For the artist—and aren't we all to some extent and at different levels artists at some moment or another in our lives?—the creative act means the total commitment of his/her whole being.

The experience of omnivalence is a cornerstone of creativity and inner leadership.

Yet, is there a form of creativity and leadership which would enable us to confront in a better way the challenging problems of today's societies? If so, are certain forms of creativity possible or even necessary to help us to survive, or—what is still more important—help the diversity of life to survive?

What kind of creativity and inner leadership is needed for young people to regain the conviction that they all are, and can be, creative? How can culture be prevented from becoming more and more a product-driven concern? What role does creativity and inner leadership have to play in the changing approach to the concept and environment of work, as well as in utilizing the ever-increasing leisure time that will be at our disposal? How may the self-confidence and true joy of fully participating in creative activity be rekindled in young people?

Does part of the answer lie in catalyzing social creativity? Drawing on the deepest and most potent of its sources at our disposal: science, art, religion and education?

Certainly a cornerstone and pivotal part of the answer can be found in the shift in perspective, the turning within that I describe above.

This turning within which triggers new modes and wide expanses of thinking is, indeed, shifting our thinking inward. By the same token, you are shifting and expanding your consciousness as well reaching out to the consciousness of other people, that of the planet and that of the universe by the sheer law of affinity and resonance. And you are starting to grasp the unity behind all things, as well as the love, harmony and beauty of that unity—which is the essence of the Sufi approach.

Let me end with a poem by *Lao Tzu* (sixth century B.C.) and two quotes of wisdom, the first by *Gandhi* and the second by the Hungarian poet *Sándor Weöres*:

> There is no need to run outside
> For better seeing,
> Nor to peer from a window. Rather abide
> At the center of your being;
> For the moment you leave it, the less you learn.
> Search your heart and see
> If he is wise who takes each turn:
> The way to do is to be.
>
> Lao Tzu (1994, p. 91)

Gandhi: "Our real power does not lie in our ability to change the world but in that of being able to recreate ourselves". (Gandhi 2008, pp. 12–13)

Sándor Weöres: "To become a genuine human being, the essence of who you are, turn within, explore your depths and transilluminate yourself, and, then, radiate your inner light on those around you and your environment". (Weöres 2011, p. 85) (translated by the author)

References

Bohm, D. 1985. *Unfolding meaning: A weekend of dialogue with David Bohm*. Gloucestershire: Foundation House.

Brabandere, L. de. 2005. *The forgotten half of change—Achieving greater creativity through changes in perception*. Chicago: Dearborn Trade Publishing.

Bradbury, R. 1953. *Fahrenheit 451*. New York: Ballentine Books.

Gandhi, M. 2008. *The words of Gandhi*. New York: Newmarket Press.

Gobillot, E. 2011. *Leadershift—Reinventing leadership for the age of mass collaboration*. London: KoganPage.

Hundertwasser, F. 2004. *Beautiful paths, thoughts on art & life—Writings 1943–1999*. Munich.

Kennedy, J. F. 1964. *Containing the public messages, speeches and statements of the president*. Washington, DC: U.S. Government Publishing Office.

Khan, H. I. 1978. *Sufi Order Publications*. New York: New Lebanon.

Khan, P. V. I. 1982. *Introducing spirituality into counseling and therapy*. Sante Fe, New Mexico: Omega Publishing.

Lao Tzu. 1994. *The way and its power: Lao Tzu's Tao Te Ching and its place in Chinese thought*. (UNESCO Collection of Representative Works, trans. by Waley, Arthur). New York: Grove Press.

Rogers, C. R. 1977. *On becoming a person: A therapist's view of psychotherapy*. London: Constable.

Saint-Exupéry de A. 1984. *The wisdom of sands*. Chicago: University of Chicago Press.

Weöres S. 2011. *Eternal Moment—Selected Poems*. Budapest: Corvina Kiadó.

András László is Founding President and CEO of GlobalVisioning.net (www.globalvisioning.net). As a Global Cultural Creative Emergence Company, GlobalVisioning.net is to create a Global Spiritual Platform of Sustainable Excellence. László was Vice President and General Manager of the Club of Budapest Foundation. He is a Founding Fellow of the Global Collaborative Alliance, Director International Affairs of the Hungarian Institute for Strategic Research, Full Member of the European Academy of Sciences and Arts, and Full Member of the Club of Rome EU Chapter. He has been a professional simultaneous conference interpreter as well as a free-lance journalist for the last 20 years, and a musician (violinist).

Deep Ecology and Personal Responsibility

Knut J. Ims

Inspired by Arne Næss' writings on ecology, this paper explores his viewpoint about the necessity of a Deep Ecological approach and the need to take a personal ecosophical position. According to Næss, deep ecology is proceeding in two directions; it is trying to bring about a change and it is trying to develop an alternative philosophy called "eco philosophy". While eco philosophy is a descriptive science without fundamental postulates about values, ecosophy includes the development and eliciting of norms. Arne Næss invites everybody to work out their own individual ecosophy, by which he means a philosophy of ecological harmony. This kind of *sophia* or wisdom is normative and contains value priorities as well as norms.

Concerning contributing to practical change, Arne Næss walked his talk and participated directly in society's value struggles. He learned from ecology as a science, but as a philosopher he was very aware of the limits of science and understood that ecology and ecologism could never substitute for philosophical analysis. Any change has political implications and Næss actively tried to fight against the "depolitization" of ecological science from the 1970s onwards. Science in itself cannot provide the ultimate answers to ethical questions about how we should live. In the same manner that we should be aware of the danger of ecologism, we should also see economism as a threat to our Western society.

In the spirit of Arne Næss this paper attempts to broaden our understanding of a deep ecological approach, and emphasizes the need to develop a personal worldview. Today there is an urgent need to create an alternative pathway for the Western World to develop along (see for example, Zsolnai 2011). *The Living Planet Report* (2008) presents strong evidence that the exploitation of natural resources and the level of consumption in Western countries have lead to a situation of huge ecological overshoot. This is one reason why the thinking promoted by deep ecology has garnered many supporters over the last decades. There are now green movements, foundations in different parts of the world, educational programs, discussion groups and even a church of deep ecology.

K. J. Ims (✉)
Norwegian School of Economics, Bergen, Norway
e-mail: Knut.Ims@nhh.no

L. Zsolnai (ed.), *The Spiritual Dimension of Business Ethics and Sustainability Management*, DOI 10.1007/978-3-319-11677-8_5

1 Deep Ecology and its Inspirational Sources

Deep Ecology was not invented by the Norwegian philosopher Arne Næss. As Glasser writes, Næss has given old ideas "a name, structure, and a philosopher's insight to a minority tradition, a quiescent utopian social movement that has been with humans since at least Buddha, Chuang Tzu, Asoka, Saint Francis of Assisi, Thoreau, John Muir, and Aldo Leopold" (Glasser 2011, p. 58).

Arne Næss himself referred to some of the inspirational sources behind deep ecology by using the metaphor "The deep ecological tree", by which he meant a tree with long and strong roots and different branches consisting of ideas from Hinduism, Confucius and Buddha on the one hand, and Aristotle, Heidegger and Spinoza on the other.

In accordance with a central Aristotelian view, every living creature has a *telos*, a goal, and can be expected to realize itself and to flourish. Gandhi, by giving concrete examples of how to live a simple, but a rich and non-violent life, was not the least significant source of inspiration for Arne Næss. Gandhi's metaphysics represents the concept of oneness. It implies that everything is interrelated and that only a holistic worldview is adequate. Such a worldview naturally leads to the undertaking of practical action on a non-violent basis.

Spinoza (1632–1677) is another major source of inspiration for Arne Næss. In his book on Spinoza, Arne Næss (1999b) writes that many of his intellectual friends have great respect for Spinoza, but wonder how it is possible for anybody to take Spinoza's metaphysics seriously today. Spinoza claims that we have an ability to identify with others and thereby come close to all kinds of life, assuming *biophilia* to be a central human attribute. Contemporary ecologist Wilson writes that biophilia is "the inborn affinity human beings have for other forms of life, an affinity evoked, according to circumstance, by pleasure, or a sense of security, or awe, or even fascination blended with revulsion" (Wilson 1994, p. 360 cited in Glasser 2011, p. 53).

As a philosopher, Arne Næss was always asking fundamental questions like "how can we deal with the essential ideological struggle of our time?", "how can we care for future generations?", "what is the place of humans in nature?", and "which technology and consumption pattern are appropriate?" It was in order to get to the roots of such problems that deep ecology was named and structured.

2 Shallow Versus Deep Ecology

In the early 1970s Arne Næss criticized dominant environmental thinking for only being concerned with short term problems and measures. What we need, Næss said, is a *long-term, deeper approach.* Næss argued that we need to distinguish between deep and shallow approaches to environmental problems. *Shallow ecology* represents a technocratic attitude to pollution control and resource depletion by supporting rules like 'the polluter pays' and assuming that symptoms can be treated through technological fixes. It follows that ecological footprints will be smaller and thus, after some time, there will be no serious ecological problems at all. Such a shallow approach assumes an anthropocentric worldview in which human affluence is vital. As a consequence, levels and patterns of production and consumption are not challenged.

In contrast, *deep ecology* assumes a relational, total field perspective that fits into a non-reductionist, non-anthropocentric worldview. It focuses on the underlying causes; *the roots of the problems*. Accordingly, the cure is more radical. The prescribed medicine is to change the basic ideological structure, which ultimately means changing how we as humans regard ourselves. Thus deep ecology redefines the very notion of self as a subject—and opens it up to transformation into an eco-Self.

To sum up, deep ecology is both a philosophical approach and a campaigning platform. It is a process of reflection leading to action.

3 Arne Næss—the Man and Philosopher

Arne Næss was born in 1912 and studied in Oslo, Paris and Vienna. At the age of 24 he took a doctoral degree and at 28 became the first full Professor of Philosophy at the University of Oslo. For 30 years he served as chair of the philosophy department in Oslo. At the age of 58 he retired and became an environmental activist and was directly involved in many demonstrations aimed at protecting the wilderness, rivers and waterfalls in Norway. At 92 years of age Arne Næss was involved in a demonstration in front of the Norwegian parliament.

Uebel (2011) emphasizes that Arne Næss was 'a kind of metaphysician' who did not exclude 'conceptual poetry' from metaphysics. Næss leaped from Logical Empiricism in his early career to Ecophilosophy as a mature man. As Uebel argues, Naess wanted philosophy to be open to the "questions of life". In Naess's own words, he was "looking for wisdom through the study known as ecophilosophy". But Næss did not stop with philosophy but strived for ecosophy—"a total view inspired in part by the science of ecology and the activities of the deep ecological movement". Ubels conclusion is that Naess remained remarkably consistent throughout his philosophical career. He succeeded in carrying out his early plan to develop an *a-metaphysical* philosophy. This attitude accords with the approach of Wittgenstein: philosophy should not be regarded as a system of distinguished propositions but primarily as the process of clarifying meaning (Uebel 2011, p. 101).

In his work on ecosophy, Næss emphasizes that we need to undertake normative evaluations and that objective science cannot provide us with the principles for action. "Without an ecosophy, ecology can provide no principles for acting, no motive for political and individual efforts" (Naess 1989, p. 41). Also, experts need norms in order to act. For Næss this awareness lead to the need to "take greater account of the pragmatic dimension of his philosophy than is common for academics. Exercising powers of persuasion over more than one's colleagues (…) became an essential part of his task" (Uebel 2011, p. 79).

In Næss' many works it is not difficult to see the underlying reasoning that explains his behavior as an activist engaged in civil disobedience. For Næss it was clear that one should not reduce himself or herself to only being an instrument for other people's projects. Accepting that we daily decide between conflicting considerations and interests, Næss emphasized that working for a "more ecologically responsible personal and societal lifestyle is not only the ecologist's job. Nor the philosopher's. We should all do it together, as generalists rather than as specialists. We all have a certain amount of practice in choosing courses of action". Also, Næss stresses that "to act in an organization merely as a functionary is to renounce human dignity in favor of the anonymity of the automaton (Naess 1989, pp. 44–45).

4 Ecosophy—The Wisdom of Living Systems

Ecophilosophy may be defined as "the utilization of basic concepts from the science of ecology—such as complexity, diversity and symbiosis—to clarify the places of our species within nature through the process of working out a total view" (Naess 1989, p. 3). The central ecological principles are *unity* and *diversity*. This means that differences in perspectives should not be lost in unity. The environmental movement is stronger if a concise set of principles are derived from a variety of world-views. This position reflects a pluralist view.

The word *Ecosophy* comes from the Greek terms *oikos* and *sofia* which mean "house" and "wisdom", respectively. In ecology, "eco" has a much wider meaning than housekeeping and can be interpreted as caring for the earth (Næss 1999a, pp. 18–20). Ecosophy is a personal philosophy; a reasoning process that may lead to something like the platform of deep ecology. Given that it is a personal philosophy, there may be many other Ecosophies (A, B, C, D…X). Arne Næss' own system of reasoning is called Ecosophy T. The T symbol stands for his mountain hut, Tvergastein, but it is the personal nature of the system that is the main point (Naess 1989, p. 4). Næss acknowledges that every person may have a different, distinct and valid Ecosophy. An underlying assumption is that, as people, we pretend to act and decide on the basis of having a total overview.

5 The Four Level System

To better understand the deep ecological worldview, it may be fruitful to consider the four levels it is said to be composed of. The most basic level is the metaphysical level. The second is the platform level and the third level relates to the policy level. The fourth level refers to level of political action (Table 1).

Table 1 is a simplified form of Naess's Apron Diagram (See Drengson and Inoue 1995; Naess 1995).

Table 1 Four levels of questioning and articulation

Level I	Ultimate premises	Taoism, Christianity, ecosophy T, etc.
Level II	Platform principles movement	Peace movement, deep ecology movement, social justice movement, etc.
Level III	Policies	A, B, C, etc.
Level IV	Practical actions	W, X, Y, etc.

The logic behind the framework is that there should be a continuous process of going back and forth between the four levels to keep our understanding and practice in harmony.

On *level one,* posing deep questions and exploring ultimate premises and norms is a primary task. Having established a more or less articulate position on the first level we may move to the lower levels. This framework admits a great diversity of ultimate philosophies. We do not have to subscribe to the same ultimate ecological philosophy in order to work cooperatively. The frontier is long—and each person may develop his or her own premises.

Næss invites everybody to work out his or her *individual ecosophy* which is openly normative and contains norms, value priorities and hypotheses.

Næss' view begins with one basic norm: *Self-Realization!* This is understood as "Self-realization for all beings!" The Self that should be realized for humans is not the ego self but the larger ecological Self.

Næss focuses on the human ability to identify with a larger sense of Self. Humans naturally have this capacity. This can be observed as a cross-cultural phenomenon.

Supporting the principles of the peace movement is also a part of Næss' philosophy. But Næss stresses that social justice is not enough. We have to produce and consume less—tread more lightly and more wisely on the Earth. His motto is "Simple in means and rich in ends". This speaks for quality of life instead of standard of living, and celebrates the virtues of slowness and smallness in an age of speed and scale.

To sum up, Næss' holistic worldview negates the dominant metaphysics of our age which sees humans as essentially different from the rest of nature. Næss claims that humanity is inseparable from nature: *If we hurt nature, we hurt ourselves.*

Taking a closer look at level two, the platform level, we can explore what unites different individuals and groups in the Deep Ecology movement. The platform is typically summed up using eight points. The paper lists the eight points next and elaborates on some of the points that may be more controversial and difficult to understand.

1. The well-being and flourishing of human and nonhuman life on earth have value in themselves—independent of their utility to humans.
2. Richness and diversity of life forms are also values in themselves.
3. Humans have no right to reduce this richness and diversity except to satisfy vital human needs.
4. We need a substantial decrease in human population.

5. The present interference with the nonhuman world is excessive.
6. Policies which affect basic economic, technological, and ideological structures must be changed.
7. The ideological change is mainly that of appreciating life quality (dwelling in situations of inherent value) rather than insisting on an increasingly higher standard of living. There will be a profound awareness of the difference between big and great.
8. Those that support these points have an obligation to directly or indirectly try to implement the necessary changes, in a non-violent way.

Another great environmentalist, *Aldo Leopold*, in his book *A Sand County Almanac* writes about an experience with a wolf that triggered a total reorientation of his life and, finally, the development of an ecocentric land ethic (Leopold 1966).

The moral of Leopold's story corresponds nicely with Næss' writings, and the story goes like this: The US government had developed a rational, scientific policy for eradicating wolves from the entire US and Leopold was appointed to implement this policy. One morning, he was out with some friends on a walk in the mountains. Being hunters, they carried their rifles with them in case they got a chance to kill some wolves. Around lunchtime they sat down on a cliff. Soon they saw a group of wolves. They took up their rifles and began to shoot excitedly into the group, but with little accuracy.

> In a second we were pumping lead into the pack, but with more excitement than accuracy: how to aim a steep downhill shot is always confusing. When our rifles were empty, the old wolf was down, and a pup was dragging a leg into impassable slide-rocks.
> We reached the old wolf in time to watch a fierce green fire dying in her eyes. I realized then, and have known ever since, that there was something new to me in those eyes—something known only to her and to the mountain (…); I thought that because fewer wolves meant more deer,that no wolves would mean hunter's paradise. But after seeing the green fire die, I sensed that neither the wolf nor the mountain agreed with such a view. (Leopold 1966, p. 130)

Two decades later, a masterpiece of a book on *land ethics* by Leopold was published posthumously which stated that: "A thing is right when it tends to preserve the integrity, stability, and beauty of the biotic community. It is wrong when it tends otherwise" (Leopold 1966, p. 240).

Supporters of deep ecology stress the importance of such spontaneous experience (seeing the fierce green fire dying in the wolf's eyes). A key aspect of these experiences is the perception of gestalt—of networks of relationships. In such gestalt we feel a strong sense of *wide identification*, an increased sense of empathy and a natural inclination to protect non-human life.

Point seven underscores that we have to stop the thoughtless consumption of material goods that has a devastating effect on the environment and does not lead to increasing happiness. "More is always better" is a tragic fallacy believed by much of the Western world (Zsolnai 2011). This point involves the need to shift from a welfare approach (measured by economic growth in GDP) to a quality of life and happiness approach (for a review of this literature, see Ims 2013). Often, more is less and less is more.

Barbara McClintoc, who won a Nobel Prize for her work in biology, writes that "one must have the time to look, the patience to hear what the material has to say to you, the openness to let it come to you. Above all one must have a feeling for the organism" (see Hallen 1995, p. 207). Mclintoc says: "no two plants are exactly alike. They are all different (...) I know them intimately and I find it a great pleasure to know them" (Keller 1983 cited in Hallen 1995, p. 207). Hallen emphasizes that this is an example of embracing the world, which is quite different from desiring to conquer it. While Francis Bacon expressed the desire to dominate, manipulate and control, or to "put nature on the rack and torture her", Hallen and McClintoc's love of plants represents an alternative tradition of science which we need to understand and emulate. This view stands in strong contrast to a science based on detachment and the division between subject and object, between self and the other.

Human happiness does not arise from consuming the maximum amount of material goods. We can learn from Buddhist economics that consumption should be guided by the 'Middle Way'. "The effective path lies between the extremes of hedonistic self-indulgence and sensual pleasure, and self-mortification or asceticism" (see Daniels 2011, p. 46).

A crucial question to ask is this: how do the supporters of these eight points contribute to the necessary and radical changes? Transition strategies are vital for reducing our ecological impact. In the field of business, organizations like *The Natural Step* are working within such a framework.

So what are the main critiques of these ideas? Criticism has emerged concerning androcentrism (or patriarchal values), population control, and accusations of misanthropy when animals and humans are presumed to have an equal footing (see Zimmerman 1995).

Concerning the critique of androcentrism, feminists may maintain that the central dichotomy is not the 'human—non human' relationship, but the 'male—female' one. The dichotomy between humans—non humans results in a lack of attention being given to such issues as the domination of humans by other humans and the exploitation of women by men. The deep ecology movement should give priority to deeply questioning the validity of socio-cultural institutions that are based on traditional patriarchal values.

6 Strategies for Changes

There are many ways to work for change. We can:

- Live according to ecological ideals.
- Encourage compromise between the present and the ideal state, nurturing the coming into existence of a realistic transition path to a new society.
- Try to change the system directly, for example via political action and acts of eco-sabotage.

Arne Næss wrote about the important balance between actively participating in direct actions and working more from day-to-day with less spectacular means. A main aspect with any actions is to attract the attention of the public. While an action is part of a campaign and a campaign is part of a movement like deep ecology, it is necessary to adhere to strict non- violent actions. The possibility of success is highly dependent on the level of nonviolence in the actions, campaign, and movements. In this context it is essential to contact the opponent before a direct action. Næss refers to the Alta (trying save the river) demonstrations (1980–1981) in arctic Norway where the powerful opponents were continually contacted for at least nine years of the campaign, and during the direct actions the opponents were treated with coffee and invited to discuss the action in order to avoid misunderstanding. Næss had learnt from the Gandhian approach to 'maximise contact with your opponent!' This strategy fits well to being a philosopher who always has to argue for his positions. Næss admitted that he had learned much from ecology as a science. But he emphasized that there are *limits to science.*

Ecology can never be a substitute for philosophical analysis. Any change has political implications, and, in Næss' words, "we need to fight against depolitization". Governments often hire experts from natural sciences like physics, chemistry, mineralogy to avoid questions of how to change our societies in the direction of sane ecopolitics. Ecology as a science, concerned with facts and logic alone, cannot answer ethical questions about how we should live. We have to go to ethics, the practical branch of philosophy, to get answers to such questions.

Ingebrigtsen and Jakobsen (2004, pp. 50–52) argue that value pluralism is important and that it is crucial to see different sectors living in reciprocal tension with each other in order not to be colonized by economic models, economic concepts and fixed interpretations. They differentiate between economy, nature and culture, seeing these sectors as being involved in a simultaneous, complex interplay, although they acknowledge that the sectors have characteristic traits that make them genuinely different. One implication of this is that the interplay between economy and culture has to be based on their characteristic value differences and that economic values should not be mixed with, or displace cultural values. The same relationship is important for the interplay between economy and nature.

From an ecosophical perspective it is important that neither economy nor ecology is accepted as the dominant view. From an ecosophical perspective, economy and ecology should be *means* within an ecosophical framework. In the following illustration such a framework is presented as a hierarchy (Fig. 1).

In the spirit of deep ecology, humans are equal in value to other sentient beings. However, this position does not imply that we as humans can drop our burden of responsibility. As human beings we have immense power in the world and therefore far-reaching responsibility (*Noblesse oblige!*). The existing pattern of production and consumption is disastrous and unsustainable. We need a radical change of mindset. As Glasser maintains, "Nature can—and will—continue to be viewed in instrumental terms as a resource and waste sink, but its use must become framed by the larger context of nature as a partner, friend, inspiration, mentor, and final

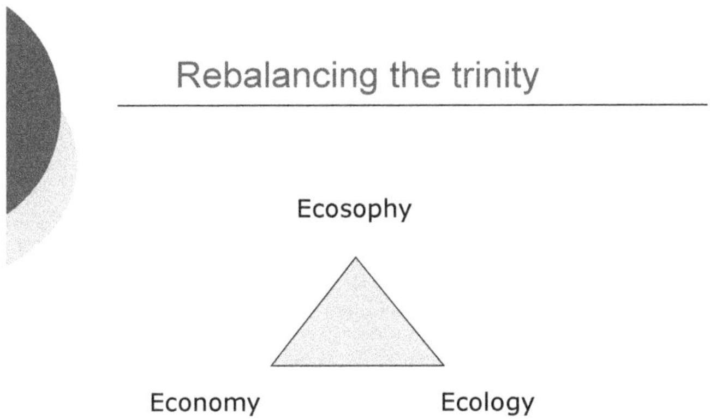

Fig. 1 The hierarchy of ecosophy, economy, and ecology

arbiter" (Glasser 2011, p. 55). We therefore need to ask a deep question: what is good for the community of life on Earth?

The story of Leopold's shooting of the wolfs shows that there is *hope* for a change in mindset. Sometimes radical changes occur sooner than anybody thinks possible.

In Norway there are strong instrumental and institutional forces that try to reduce political issues to problems to be solved by experts. Vital issues are often not debated in an open and democratic manner. This practice implies that specialists follow established norms and do not take ecological issues seriously. Often the prevalent norms satisfy rather narrow economic conditions as if the cheapest solution were always the best. Experts from different fields can make important contributions as regards clarifying the means that is appropriate for achieving a given goal. However, economists and ecologists should be engaged and take explicit positions about controversial issues in society and not let professional politicians debate the fate of society alone. Experts are not better equipped than non-experts to decide which values and norms should be part of the value system of society. Experts should participate in the process of deciding what goals and priorities local communities have, as well as society as a whole.

Ecosophy is a personal position—a personal total worldview. Everybody should work out a worldview and act in accordance with it. It is not possible to articulate a complete 'total view', but by clarifying fundamental norms and values one will be better able to see and respond to the challenges of life in the ecosphere. As Arne Næss wrote, "the ecopolitical frontier is immensely long…" (Naess 1989, pp. 161–162). It follows that there is work for everyone to change the system by using his or her own capabilities.

References

Daniels, P. 2011. Buddhism and sustainable consumption. In *Ethical principles and economic transformation—A Buddhist approach.* Vol. 33, ed. L. Zsolnai, 35–60. Springer, Series. Issues in Business Ethics.

Drengson, A., and Y. Inoue. 1995. Introduction. In *The deep ecology movement: An introductory anthology,* ed. A. Drengson and Y. Inoue, xvii–xxviii. North Atlantic Books, Berkeley.
Glasser, H. 2011. Naess's deep ecology: Implications for the human prospect and challenges for the future. *Inquiry* 54 (1): 52–77.
Hallen, P. 1995. Making peace with nature: Why ecology needs feminism. In *The deep ecology movement, an introductory anthology,* ed. A. Drengson and Y. Inoue, 198–218. Berkeley: North Atlantic Books.
Ims, K. 2013. From welfare to well-being and happiness. In *Handbook of business ethics. Ethics in the new economy,* ed. L. Zsolnai, 217–247. Oxford: Peter Lang Academic Publishers.
Ingebrigtsen, S., and O. Jakobsen. 2004. *Økonomi, natur og kultur*. Oslo: Abstrakt forlag.
Leopold, A. 1966. *A Sand County almanac. with other essays on conservation from Round River*. New York: Oxford University Press.
Naess, A. 1989. *Ecology, community and lifestyle. Outline of an ecosophy*. (trans. and revised by D. Rothenberg). Cambridge: Cambridge University Press.
Naess, A. 1995. The apron diagram. In *The deep ecology movement: An introductory anthology,* ed. A. Drengson and Y. Inoue, 10–12. Berkeley: North Atlantic Books.
Næss, A. 1999a. *Økologi, samfunn og livsstil*. 5th. revised ed. Oslo: Boklubbens Dagens Bøker.
Næss, A. 1999b. *Det frie menneske. En innføring i Spinozas filosofi*. Oslo: Kagge Forlag.
Uebel, T. 2011. A kind of metaphysician: Arne Naess from logical empiricism to ecophilosophy. *Inquiry* 54 (1): 78–109.
Zimmerman, M.E. 1995. Feminism, deep ecology, and environmental ethics. In *The deep ecology movement, an introductory anthology,* ed. A. Drengson and Y. Inoue, 169–197. Berkeley: North Atlantic Books.
Zsolnai, L. 2011. Why Buddhist economics? In *Ethical principles and economic transformation—A Buddhist approach*. Vol. 33, ed. L. Zsolnai, 3–20. Springer, Series. Issues in Business Ethics.

Knut J. Ims is professor in Business Ethics at the Norwegian School of Economics (NHH) in Bergen, Norway. He received his PhD from the School of Economics and Legal Sciences, Gothenburg University, Sweden. He has taught courses in ethics (Ethical Action—Individual, Organization and Society; and, Business Strategy and Business Ethics) for more than a decade. He has also taught PhD courses in system development and information and management. He is a member of the Business Ethics Interfaculty Group of the Community of European Management Schools (CEMS), and is chairman of the board of the Centre of Ethics and Economics at NHH.

Buddhist Spiritual Orientation to Nature and Sustainability

Gabor Kovacs

Economic growth is massively idolized in our modern world. Nothing is too much for the achievement of this goal, and its long-term effects are frequently ignored. The paper explores the Buddhist spiritual orientation to nature and interprets sustainability from the Buddhist point of view. It describes Buddhist virtues that could be helpful in solving environmental problems via the promotion of frugal livelihoods.

1 The Causes of Today's Environmental Crisis

Western materialism has gradually raised human beings 'out' of nature and made them the dominant planetary species based on their anthropocentric worldview. Today's societies are detached from nature. The consequence of having an exclusively anthropocentric worldview is the emergence of a growth-oriented, globalized economy. Consumption-based and greed-driven human lifestyles are resulting in overuse of the resources of the planet and pollution of the environment to an ever-greater extent. The consequences of these dismal global processes are deforestation, soil degradation, loss of biodiversity, the extinction of species, resource depletion and climate change—just to mention the most significant environmental problems.[1] Diamond (2005) uses a five-point framework to explain the contributory factors to social collapses across history and has arrived at the conclusion that the fifth factor—society's response to environmental problems—is always a significant factor in whether societies survive.

[1] The growth-oriented, globalized economy also results in many negative social consequences: ever-increasing levels of inequality and social insecurity, massive poverty and the centralization of economic and political power, for example.

G. Kovacs (✉)
Corvinus University of Budapest, Budapest, Hungary
e-mail: mr.malasa@gmail.com

L. Zsolnai (ed.), *The Spiritual Dimension of Business Ethics and Sustainability Management*, DOI 10.1007/978-3-319-11677-8_6

As Capra (1982) stated, social and environmental crises have a major spiritual dimension because the root of the problem is the heart of human spirituality. Bearing this in mind, an investigation of spiritual traditions—in this case, Buddhism—represents a fruitful approach to examining the current situation from another standpoint.

2 The Teachings of Buddhism

The teaching of Buddhism deals with suffering and the cessation of suffering. Its central doctrine is the Four Noble Truths, which were articulated in Buddha's first sermon and have been enunciated numerous times since: "*I teach only two things, the fact of suffering and the possibility of escape from suffering*" (SN 22.86, MN 22).

The First Noble Truth states that the true nature of sentient beings is suffering. The Buddha declared that numerous events in life are painful—birth, ageing, sickness and death. To be together with those who are undesired or to be away from loved ones is also painful. Suffering is also present when one's desires are not fulfilled—this feeling is very prevalent in modern consumption-based lifestyles.

The Second Noble Truth deals with the cause of suffering. It asserts that its origin is in the widespread human characteristic of unchecked craving for the ephemeral phenomena of the impermanent world. Aging, sickness and death—the accompaniments of life—are accompanied by suffering if one clings to youth, health and life. Transience is an inherent characteristic of every phenomenon, as everything that once was born or constructed will die or disintegrate. Misery exists because of our attachment to certain states of impermanent feelings and uncontrolled greed for ephemeral things—especially goods produced by the consumption-based society of our times.[2]

The Third Noble Truth declares that suffering can be ceased via the elimination of attachment and greed. This assertion leads to the final truth, The Fourth Noble Truth, which states that the tool for this is The Noble Eightfold Path.[3] This is lifestyle advice, which refers to 'right behavior' that leads to the ultimate goal of Buddhists; the final cessation of suffering—the overcoming of greed, hatred and delusion, the three root causes of suffering (SN 56.11, Batchelor 1992).

According to the teachings, greed, craving or uncontrolled desire are the main causes of personal suffering, but this can be extended beyond the personal dimension. Members of a growth-oriented consumer society direct their limitless desires towards goods, wealth and power and these aspirations have significant and

[2] Greed is strengthened by sophisticated marketing tools which are designed to maximize consumption, but the operation of these forces is the subject of another paper.

[3] The Noble Eightfold Path is comprised of eight parts (in three groups). These are: (i) Right View and (ii) Right Decision (the 'wisdom' group); (iii) Right Speech, (iv) Right Action, (v) Right Livelihood (the 'virtue' group); (vi) Right Effort, (vii) Right Mindfulness and (viii) Right Concentration ('spiritual practice')

immediate consequences on the environment. Humanity's greed is one of the foremost spiritual causes of environmental degradation and destruction (Batchelor 1992; Schmithausen 1997; Silva 2000; Dharmakosajarn 2011).

3 The Place of Humankind in Nature—Buddhist Cosmology

The tradition of Buddhism offers a different view of the role and place of humankind in nature, since it denies human dominance over other species. It has a moderately anthropocentric worldview as it acknowledges human superiority over other beings, but this is based on their capacity for moral and spiritual development (Harvey 2000; Sponsel and Natadecha-Sponsel 2001; Pragati 2008).[4] This view is not the same as 'stewardship', or 'dominion', but it is identical with the spirit of '*noblesse oblige*' (Schumacher 1973).

Rather than dividing the world into the realms of 'human' and 'nature', the Buddhist perspective involves a more appropriate distinction between sentient beings (humans and animals) and the non-sentient environment, where sentience is defined as being the ability to experience suffering.[5] Using this distinction, plants are part of the non-sentient portion of the world, but there is some ambiguity here (Harvey 2000; Silva 2009). Since the Vedic times—from the 16th to 5th centuries B.C.—Indian people believed that special gods (*devata*) or spirits (*yakkha*) resided in plants, especially in old trees (Silva 1987; Kalupahana 2009). This tradition did not violate the belief system of Buddhism. Furthermore, it was a suitable way of transforming a part of the non-sentient flora into the sentient realm of the world, as gods and spirits are also able to experience sensual feelings. Secondly, there are Buddhist texts that state that harming seed and plant life is strictly prohibited (DN 1, Vin IV 34, Vin I 137, Vin IV 205–206).

Humans are part of the community of sentient beings—all of them share the same characteristics of suffering (birth, aging, sickness and death). The spiritual potential of humans—the fact that they are potentially able to overcome greed, hatred and delusion—makes them more valued than other species, but that potential should be expressed through compassionate behavior and sympathy towards other beings and nature. Humans should not dominate, but act as neighbors to less intelligent and less fortunate sentient beings (Harvey 2000). Thus human superiority does not act in opposition to maintaining the integrity of the natural environment.

---

[4] Human beings have the utmost importance according their ability to become enlightened—the final aim of Buddhism –, in contrast to the other planes of existence in the Buddhist cosmology: (i) the realm of hells, (ii) animals, (iii) hungry ghosts, (iv) human beings, (v) semi-gods, and (vi) gods.

[5] The Buddhist distinction is very similar to that used by Peter Singer, who defined 'sentience' as the criterion for moral relevance, the key quality of ethical '*sentientism*'.

4 The Role of Nature in the Early Buddhist Tradition

In the following section the paper will discuss the role of nature in the early Buddhist tradition. Early Buddhism is the traditional teaching of the Buddha, which is referred to as Hinayana Buddhism.[6] This distinction is necessary, as the role and the interpretation of nature in Buddhism changed as it spread through Asian cultures. The cause of this transformation is syncretism, and it implies that Buddhism's tenets became merged with the local traditions it encountered. Syncretism has resulted in several different interpretations of the role of nature in Buddhism.[7]

By examining the sacred scriptures of Hinayana Buddhism, one can find that flourishing natural habitats, forests and groves were part of the life of Gautama Siddhartha, the Buddha. Furthermore, every major event in his life was related to trees: he was born in Lumbíni under a sal-tree in a small park in North-East India; he attained his enlightenment under the bodhi-tree; his first dwelling place was Veluvana, a bamboo grove; he traveled during his whole life through the wilderness, taught in forest-groves and died beneath a pair of sal-trees (Harvey 2000; Dharmakosajarn 2011).

A part of the Buddhist Canon contains the Jatakas, the past-life stories of the Buddha, in which more than five hundred tales testify to the value of nature. They highlight the importance of intact animal and plant life and are important in the environmental education of young children (Pragati 2009; Dharmakosajarn 2011).

There are texts about the Buddha's appreciation of beautiful places: "I saw some delightful countryside, with an inspiring forest grove, a clear-flowing river with fine, delightful banks, and villages for alms-going on all sides… This is just right for the striving of a clansman intent on striving" (MN 12, MN 36). The reason for his enthusiasm is that an untouched environment contributes to the overcoming of the five hindrances of enlightenment to the highest degree (MN 38). These places are calm as they don't generate any sensual pleasures; one can't find anything else to do but contemplate, so they provide experimental ground for the basic teachings (Kalupahana 2009).[8]

Forest-groves are not just ideal places for meditation but trees are especially fascinating for the Eastern mind as symbols of spiritual freedom (Silva 1987; Harvey 2000). This is in stark contrast to the modern approach which looks upon the environment as something primarily for human consumption (Kalupahana 2009).

[6] Nowadays Hinayana Buddhism is a living tradition in the countries of South-East Asia—Thailand, Vietnam, Mianmar, etc.—and in Ceylon.

[7] Significant differences exist between Buddhist practices in Hinayana countries and, for example, Japan, where the native tradition of Shintoism, an ancient animism, had a remarkable impact on the interpretation of the role of nature in Buddhism.

[8] Although calm forest groves are the best places for Buddhist meditation practice, they may be substituted for by other locations. The Buddha himself, inter alia, advocated meditation in cemeteries, where people can contemplate death and the impermanence of the human body (DN 22, MN 13). Modern Hinayana-teachers also draw attention to the ever-increasing distractions of our age, and confirm that Buddhist meditation can be practiced anywhere regardless of the place. Success does not depend on circumstances (Sumedho 1991).

5 Buddhist Spiritual Orientation to Nature

In the time of the Buddha (the fifth century BC), ecosystems were still intact and there were no species threatened with extinction. Pollution was not as significant as it is today and environmental degradation wasn't so dramatic. No significant environmental crises were known about, or heard of. This is why the Buddha himself did not pay much attention to environmental affairs except for occasional cases when unequivocal harm to the environment occurred and casual regulations were necessary.[9] His motivation for addressing these incidents was twofold. He encouraged his followers to extend the spirit of non-harming towards non-sentient nature to let natural beings exist in their original state, and secondly, he wanted to maintain the renown of his community among the laity through adhering to environmentally sound precepts.

Numerous authors confirm the fact that early Buddhism does not contain a definition of 'nature', nor (in spite of the former examples) award positive value to nature in itself (Schmithausen 1997; Pragati 2008; James 2009). Nevertheless, Buddhism is not at odds with environmental concerns. It does not seek to motivate efforts to preserve or restore nature, nor to transform or subjugate it. The best one can do is to leave it as it is, with the least possible intervention.

The reason for the seeming indifference towards nature at the birth of Buddhism stems from two sources: the lack of significant environmental pollution at the time and the Buddha's pragmatic emphasis on personal suffering that had nothing to do with the environment, although Buddhists do believe that a calm milieu can facilitate one's own spiritual growth to a great extent (Schmithausen 1997). Buddha had no interest in developing a theory of nature so he strictly refused to answer questions which did not directly bear on human suffering (Silva 1987; James 2009).

This attitude can contribute only to a de facto environmentalism, and the fact that nature has no positive value in itself prevents the establishment of a Buddhist environmental ethics. As Harris notes, the minimum demanded from an authentic environmental ethics is that it has to be able to construe causation in such a way that goal-oriented activity makes sense. In opposition to this, according to Buddhism everything is impermanent and the universe lacks a sense of purpose (1094). Schmithausen (1997) and James (2009) also confirm that early Buddhist teachings can not provide a basis for ecological ethics without radical reinterpretation, if we define ecological ethics as something that gives positive value to intact nature and biodiversity in their own right.

---

[9] There are numerous cases in the Buddhist scriptures which tell of how the Buddha himself occasionally dealt with some cases related to environmental preservation. He promulgated a rule against going on journeys during the rainy season because of possible injury to worms and insects that would come to the surface (Vin I 137). He prevented monks from digging the ground and drinking unfiltered water for the same reasons (Vin IV 125). There are monastic rules which prevent monks from injuring plant life or felling trees (Vin IV 34), destroying growing crops and grasses by trampling (Vin I 137), and polluting green grass and water with saliva, urine and faeces (Vin IV 205–206).

Despite these facts, the teaching of Buddhism is often considered eco-friendly, but not because of what it says about nature, but because of what it says about the spirit of human beings (James 2009; Dharmakosajarn 2011). There is no contradiction between the lack of teachings that attribute positive value to nature and the inherently environmentally-friendly daily practice of the tradition. Schmithausen (1997) calls this "*a passive ecological attitude*" which emerges as a byproduct of Buddhist livelihood.

Buddhist teachings place us into a closely-related, interdependent relationship with our world; a view shared by modern science that agrees that society is inherently embedded in its environment. In this respect, the practice of Buddhist virtues has important implications for interacting with nature and is in line with modern environmentalism. Pragati defines this approach as "*Buddhist environmental virtue ethics*" (Pragati 2008, p. 113). This focuses on the character of the individual: morally correct conduct is evaluated through the virtues of the actor. Hinayana texts of Buddhism contain several environment-related stories and emphasize the virtues of non-harming, compassion, loving kindness, gratitude, contentment, simplicity, wisdom and mindfulness towards the sentient and non-sentient components of the environment. By practicing the virtues represented in the morals of the texts, one is automatically attuned to environmental concerns. Leading a noble life is identical to developing oneself in Buddhist virtues.[10]

The foremost virtue in regard to the environment is 'non-harming', the first of the five Buddhist precepts. This represents an approach to minimizing impact that prohibits the destruction of life. It is not just about prohibiting certain attitudes, but about encouraging active participation in doing good. In a deeper sense, it means the rejection of violence and the cultivation of positive values of love and compassion towards all sentient beings (MN 38, Saddhatissa 1970). Holding non-harming attitudes towards the environment has been interpreted to be equivalent to the preservation of biodiversity (Silva 2009) and having a non-violent attitude towards natural beings (Sponsel and Natadecha-Sponsel 2001).

Gratitude towards the non-sentient realm of nature is the another important environmental virtue. Forest-groves, trees, mountains, lakes and rivers are objects of reverence for the spirit of gratitude (Dh 188).[11] This implies the aspiration not to intervene or transform, but to revere and contemplate the intact state of the environment.

There are three closely-related virtues in Buddhist life: being content with little (DN 2, MN 3); being moderate in daily activity, like consumption of food (AN 3.17), and the practice of simplicity (DN 2). All of these virtues point towards a more sustainable and environmentally-sound livelihood. The realization of these virtues results in the fading of greed and craving and a more moderate, frugal way of consuming that has a resounding effect on the quality of nature (Pragati 2008).

Deepening wisdom may be a consequence of the practice of Buddhist mindfulness meditation. In relation to environmental concerns, wisdom means the

[10] On a special occasion the Buddha exhorted that planting trees and groves is a virtuous act (SN 1.33), and confirmed that destroying a tree is similar to betraying a friend (P 259).

[11] In the later tradition Buddhist followers started to venerate the Bodhi-tree under which Gautama Siddhartha attained his enlightenment. The tree became a place of pilgrimage.

recognition of dependent origination and the interdependent existence of natural habitats—with this the Buddha anticipated findings about ecological interrelationships.[12] Originally, the teaching of dependent origination was solely concerned with the genesis of human suffering (MN 38). Later (but still within Hinayana Buddhism) it was reinterpreted and applied to the general arising of phenomena. The universal principle of dependent origination has been described as: (i) when this is, that is; so, (ii) from the arising of this comes the arising of that (Kalupahana 2009; Silva 2009). The Aggannya Sutta deals with the genesis of the world and human societies (DN 27). Based on dependent origination it emphasizes the close relationship between human morality and the state of nature. It draws the conclusion, affirmed by other texts, that the quality of human life and the condition of the environment are closely interdependent—when moral degeneration (especially greed) becomes rampant in society, this causes adverse changes to the environment (DN 26, DN 27, AN 3.61, AN 4.70).

As stated previously, there is no direct reference to environmental thinking in early Buddhism, but each of the above-mentioned virtues represents a solid basis for a Buddhist environmental virtue ethics which can result in an eco-friendly human lifestyle.

6 The Spiritual Notion of Sustainability in Buddhism

Environmental preservation and efforts to create a state of sustainability are closely related subjects. Attitudes towards nature have significant consequences for the implementation of sustainability.

The teaching of impermanence has a crucial role in Buddhism as impermanence is a characteristic of every phenomenon. This says that everything which comes to be will decay (MN 22, SN 36.9).[13] The modern concept of sustainability was first articulated as sustainable development in the report by the Brundtland Commission (1987). As every phenomenon is subject to change, decay or cessation, sustainability in a strict sense is impossible according to Buddhism. Furthermore, clinging to permanence or to the way things are inevitably leads to suffering.

In spite of this, the heart of Buddhism calls for ongoing spiritual perfection in the form of the threefold practice, better known as the Noble Eightfold Path (SN 57.11). The observation of precepts and compliance with virtues are the first steps on this path, which leads to a non-harming, peaceful lifestyle (and implies the practice of

[12] Let us see the importance of forests in the ecological chain. According to dependent origination, if forests are cut down, drought and erosion will result as forests provide humidity, store water and fix the topsoil. Unsustainable forestry leads to environmental degradation, which leads to poverty and other social malaises.

[13] The concept of eternal impermanence has also been articulated in western philosophy and modern science. William James, Bertrand Russell, Alfred North Whitehead, Edmund Husserl, Karl Jaspers and Martin Heidegger have all contributed to the development of this notion, and Albert Einstein's theory of relativity and modern field theory have had significant influence on the development of the western concept of impermanence.

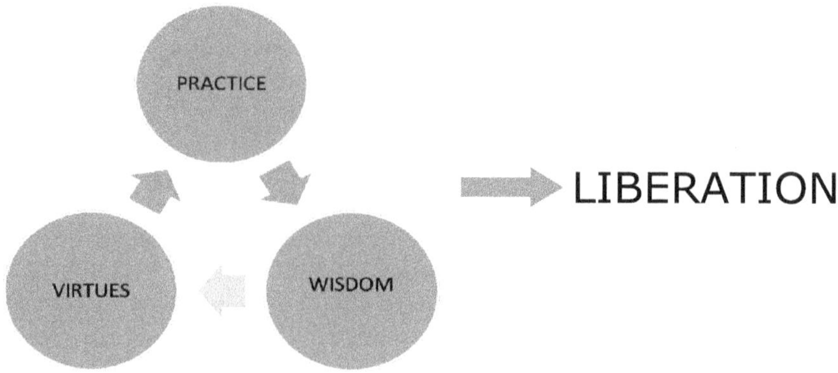

Fig. 1 Buddhist spiritual perfection

the aforementioned environmental virtue ethics). A peaceful lifestyle allows for efficient and fruitful spiritual practice which results in the tranquility of the human mind and the realization of wisdom. Wisdom leads to confidence in leading a virtuous lifestyle and results in the further strengthening of precepts and virtues. This continuous practice (living a life according to the Noble Eightfold Path) is a virtuous spiral; a path of increasing perfection that leads to the cessation of suffering, liberation and the overcoming of greed, hatred and delusion (Buddhadasa 1996; Puntasen 2007) (Fig. 1). As the Buddha asserted:

> Great becomes the fruit, great is the gain of meditation when it is fully developed by virtuous conduct; great becomes the fruit, great is the gain of wisdom when it is fully developed by meditation; utterly freed from the taints… the mind that is fully developed in wisdom (DN 16).

Buddhist spiritual perfection can be regarded as ongoing or sustained development that occurs in the form of a virtuous spiral as an emerging byproduct of the threefold practice—the 'Buddhist way of life'.

In Buddhism, sustainability is not a goal in itself, but an accompanying byproduct of a virtuous livelihood. According to Buddhism, sustainability is interpreted in the internal, spiritual realm of life, in contrast to the modern materialistic approach to economic, social and environmental sustainability.

As Schmithausen (1997) emphasizes, Buddhism, like any other spiritual tradition, cannot avoid facing the problems of modernity. If it wants to remain a living tradition, it has to supply answers to new and modern vital questions. From this point of view the survival of Buddhism will depend on its ability to respond to the significant questions of our age.[14] The challenge is to apply Buddhist principles and insights to contemporary issues which are more complex than those of the ancient world (Tilakaratne 2009).

[14] Early Buddhism had this ability at its birth as it had to compete with other Indian sects and other teachings for its own survival. This is why the practicing of wisdom and keeping of precepts were so important to monks – they had to show their virtuous livelihood to secular people to assure their material support.

Sustainability is not a subject that is directly embodied in the early Buddhist scriptures, although a place for it is allowed in the Buddhist canon that deals with 'the principles of growth' (Dharmakosajarn 2011). Here, four necessary factors are mentioned: (i) living in a favorable environment; (ii) associating with good people; (iii) directing oneself in the right way; and, (iv) doing formerly meritorious deeds (AN 4.31). This is the subject of Buddhism most closely related to the modern interpretation of sustainable development. There are several texts that deal with right livelihood (DN 31, AN 6.45, MN 143) and social stability (DN 5, DN 26), all of which arrive at the same conclusion as the texts that deal with environmental concerns: Buddhist virtues form the basis of right behavior in economic, social and environmental affairs.[15]

From the point of view of sustainability, reducing consumption-based desire through the practice of contentment, frugality and moderation is badly needed (Zsolnai 2008; Butler 2009a). This is also appreciated by early Buddhist scriptures (AN 7.2, AN 8.25, AN 8.30).[16]

Wisdom is crucial to being socially engaged, especially in Asian Buddhist countries, which ensures that Buddhism remains a living tradition (Butler 2009b). The emerging concept of 'Eco-Buddhism', or 'Green-Buddhism' is a form of engaged Buddhism that is based on the teaching of dependent origination (Tilakaratne 2009; Kwong-Cheong 2009). These forms of Buddhism aim to create a non-violent ecology (Sponsel and Natadecha-Sponsel 2001) and to conserve and restore nature through Buddhist engagement in environmental affairs.

7 References from the Pali Canon

The paper refers to Buddhist texts by their numbers, except for Vinaya texts which are referred to by their page numbers.

DN: Digha Nikaya
MN: Majjhima Nikaya
SN: Samyutta Nikaya
AN: Anguttara Nikaya
Vin: Vinaya
P: Petavatthu
Dh: Dhammapada

[15] A summary of the early scriptures that deals with economics has been written by Venerable Payutto (1994).

[16] Furthermore, Buddhist texts include one of the most ancient pieces of advice about 'recycling'. This is Buddha's pronouncement about the right use of clothing: when monks receive new robes, the old ones are not to be discarded, but be used as coverlets; when the coverlets are old, they are to be converted into mattress covers; the old mattress covers are again to be converted to rugs and old rugs into dusters (Vin II 291).

Acknowledgement The research work reported on in this paper was supported by the "TÁMOP 4.2.2/B-10/1-2010-0023" project undertaken at the Corvinus University of Budapest.

References

Batchelor, S. 1992. The sands of the ganges—Notes towards a Buddhist ecological philosophy. In *Buddhism and ecology,* ed. M. Batchelor and K. Brown, 31–39. London: Cassell Publishers.

Buddhadasa, B. 1996. *Handbook for mankind.* Bangkok: Dhammadana Foundation.

Butler, C. D. 2009a. The global environmental crisis and sustainability of civilization: Time for the Buddhist world to awaken. In *Buddhist approach to environmental crisis,* ed. K. Dhammasami, et al., 216–225. Bangkok: Mahachulalongkornrajavidyala Press.

Butler, C. D. 2009b. Without and within: Science and the middle path. *The Journal of the International Association of Buddhist Universities* 2:85–95.

Capra, F. 1982. *The turning point, science, society and the rising culture.* New York: Simon and Schuster.

Dharmakosajarn, P. 2011. *Dhamma and Environmental Preservation.* Bangkok: Mahachulalongkornrajavidyala Press.

Diamond, J. 2005. *Collapse. How societies choose to fail or survive.* London: Penguin Books.

Harris, I. 1994. Causation and telos: The problem of Buddhist environmental ethics. *Journal of Buddhist Ethics* 1:45–56.

Harvey, P. 2000. *An introduction to Buddhist ethics.* Cambridge: Cambridge University Press.

James, S. P. 2009. The foundations of Buddhist environmental ethics: A new approach. In *Buddhist approach to environmental crisis,* ed. K. Dhammasami, et al., 60–67. Bangkok: Mahachulalongkornrajavidyala Press.

Kalupahana, D. J. 2009. Buddhist approach to the environmental problem. In *Buddhist approach to environmental crisis,* ed. K. Dhammasami, et al., 1–11. Bangkok: Mahachulalongkornrajavidyala Press.

Kwong-Cheong, W. 2009. Problems with the holistic approach to Buddhist environmental ethics. In *Buddhist approach to environmental crisis,* ed. K. Dhammasami, et al., 159–184. Bangkok: Mahachulalongkornrajavidyala Press.

Our Common Future. 1987. *Report of the world commission on environment and development: Our common future.* http://www.un-documents.net/wced-ocf.htm.

Payutto, Ven.P. A. 1994. *Buddhist economics: A middle way for the market place.* http://www.buddhanet.net/cmdsg/econ.htm. Accessed 12. July 2012

Pragati, S. 2008. *Environmental ethics in Buddhism. A virtues approach.* London: Routledge.

Pragati, S. 2009. Environmental ethics in the Jātakas: Further reflections. In *Buddhist approach to environmental crisis,* ed. K. Dhammasami, et al., 129–146. Bangkok: Mahachulalongkornrajavidyala Press.

Puntasen, A. 2007. Buddhist economics as a new paradigm toward happiness. *Society and Economy* 29 (2): 181–200.

Saddhatissa, H. 1970. *Buddhist ethics: Essence of Buddhism.* London: Allen & Unwin.

Schmithausen, L. 1997. The early Buddhist tradition and ecological ethics. *Journal of Buddhist Ethics* 4:1–74.

Schumacher, E. F. 1973. *Small is beautiful: Economics as if people mattered.* London: Blond & Briggs Ltd.

Silva, D. L. 1987. The Buddhist attitude towards nature. In *Buddhist perspectives on the ecocrisis,* ed. K. Sandell, 9–19. Kandy: Buddhist Publication Society.

Silva, D. L. 2000. *The relevance of vipassana for the environmental crisis.* http://www.buddhismtoday.com/english/ecology/002-relevance.htm. Accessed 12 July 2012

Silva, P. 2009. Buddhism, environment and the human future. In *Buddhist approach to environmental crisis,* ed. K. Dhammasami, et al., 12–39. Bangkok: Mahachulalongkornrajavidyala Press.

Sponsel, L. E., and P. Natadecha-Sponsel. 2001. Nonviolent ecology: The possibilities of Buddhism. In *Buddhism and nonviolent global problem-solving,* ed. G.D. Paige and S. Gilliatt, 145–157. Honolulu: University of Hawaii Press.

Sumedho, A. 1991. *The way it is*. Hertfordshire: Amaravati Publications.

Tilakaratne, A. 2009. Emerging new trends in Buddhism and their doctrinal and organizational implications. *The Journal of the International Association of Buddhist Universities* 2:96–120.

Zsolnai, L. 2008. Buddhist economic strategy. In *Frugality—Rebalancing material and spiritual values in economic life,* ed. L. Bouckaert, H. Opdebeeck, and L. Zsolnai, 279–304. Bern: Peter Lang Academic Publishers.

Gabor Kovacs is PhD student at the Business Ethics Center of the Corvinus University of Budapest, Hungary. His research topic is spiritual value-orientations in entrepreneurship. He has background in economics and management. He received his master's degree in Buddhist studies from the Budapest Buddhist University in 2010. He was participating in a research project about the ecological value-orientation of Hungarian entrepreneurships. He is researching on Buddhist economics as well.

Sustainability and Wisdom: The Power of the Fable

Rita Ghesquière

This paper focuses on the fable as a sustainable good and presents some fables which foster the idea of sustainability.

1 The Fable as a Sustainable Good

The fable is an old literary genre/model that has its roots in popular tradition. The word 'fable' comes from the Latin *fabula* (discourse, story) and is linked to 'fari' which means speak/talk (especially by gods). The fable is a polygenetic genre. We find fables in every culture in the world and often they put forward a similar message.

Later on, the popular wisdom of anonymous authors turned into literature writing. Aesop was one of the first to give the fable its literary prestige.[1] Phaedrus

[1] The Perry Index is a widely-used index of "Aesop's Fables" or "Aesopica", the fables credited to Aesop, the story-teller who lived in ancient Greece between 620 and 560 BC. Modern scholarship takes the view that Aesop probably did not compose all the fables attributed to him; indeed, a few are known to have been in circulation before Aesop's time, while the first record we have of many others is from well over a millennium after his time. Traditionally, Aesop's fables were arranged alphabetically, which is not helpful to the reader. Perry and Rodriguez Adardos separated the Greek fables from the Latin ones by putting the Greek ones first; then they arranged each group chronologically and by source; finally they arranged the fables alphabetically within these groups.

Ben Edwin Perry (1892–1968) was a professor of classics at the University of Illinois at Urbana-Champaign from 1924 to 1960. He was the author of Studies in the Text History of the Life and Fables of Aesop and many other books. His Aesopica ("A Series of Texts Relating to Aesop Or Ascribed to Him Or Closely Connected with the Literal Tradition that Bears His Name") has become the definitive edition of all the fables attributed to Aesop, with fables arranged by their

R. Ghesquière (✉)
Catholic University of Leuven, Leuven, Belgium
e-mail: Rita.Ghesquiere@arts.kuleuven.be

L. Zsolnai (ed.), *The Spiritual Dimension of Business Ethics and Sustainability Management,* DOI 10.1007/978-3-319-11677-8_7

(15BC- AD 50), Leonardo da Vinci (1452–1519), Jean de La Fontaine (1625–1695) and the Russian Ivan Krylov (1768–1844) followed in his footsteps. Nevertheless, until now the fable, like the fairy tale, has remained a kind of common good. In the course of history the position of the fable has been controversial. Its wisdom has been acknowledged by famous scholars such as Erasmus, Fénelon and Locke who defended it as an excellent instrument for education. Luther, convinced of the didactic qualities of the genre, translated some of Aesop's fables into German. He praised the fable as a useful tool for all ages. He believed that not only simple-hearted family men but even gentlemen and rulers could take advantage of the wisdom of the fable.[2]

Other scholars such as Plato and Rousseau rejected fables as 'falsified accounts' of events that were therefore dangerous for children. Until now, the word fable, when used in a negative sense, means 'old wives' tale', or 'fib'.

During the eighteenth and nineteenth century the fable was considered excellent literature for children and was hence downgraded and almost banished to the schoolbook. Nevertheless, the classical texts of Aesop, La Fontaine and others remained part of cultural heritage and later authors often used the fable in a new and unexpected way. A good example of this is the American journalist James Thurber's (1894–1961) *Fables for our Time*, published in The New Yorker.

> Every writer … is fascinated by the fable form; it is short, concise, and can say a great deal about human life: the little flaws and foibles and vanities of man and his wife, or the larger political scene, or anything else (Jochum 1982, p. 236).

The fable is a short narrative in prose or verse and contains non-human creatures such as animals, plants or even inanimate things (mirror, ink or paper) as protagonists.[3] Human beings are less prominently featured as characters but they do make their appearance in some fables such as 'The farmer and his sons' (Aesop). Tales that include a combination of animals and human beings are more frequently found.

Fables combine merriness and wisdom. They are 'fiction' but the animal characters act as human beings. Hence their behavior and their circumstances are experienced by the audience as true, as an essential part of the experience of life. '*Die Warheit sagen von eusserlichem Leben in der Welt*' ('Tell the truth about real life in the world') is, according to Luther, the hidden power of the fable. (Leibfried 1967, p. 3) Although the stories in fables have not really happened, they should be plausible and hence convincing. The fable communicates its message twice: first through the story that brings forward the proof; second by the saying at the end that recapitulates the message in a nutshell. Because of this double argumentation structure some scholars (e.g. Lessing) consider the fable to be tendentious. Its main role is didactic. Herder stresses the active role of the reader or listener, who has to employ both imagination and reason. The reader has to deduce the practical wisdom that is contained in the story by connecting

earliest-known source. His index of fables has been used as a reference system by later authors. En.wikipedia.org/wiki/Perry_index. (2012. 06.12).

[2] "Nicht allein aber die Kinder/sondern auch die grossen Fuersten und herrn/kan nicht bas betreigen/zur Warheit/und zu ihrem nutz/denn das mas inen lasse die Narren die Warheit sagen/dieselbigen koennen sie leiden und hoeren/sonst woellen oder koennen sie/ von keinem Weisen die Warheit leiden." Martin Luthers Fabeln. W. Steinberg, 1961, pp. 86 (in Leibfried 1967, p. 3).

[3] Leonardo da Vinci wrote several fables with plants and objects as characters.

concrete events with the general rule. Reasoning by analogy ('animals behave like human beings') requires another form of intellectual competence.

To conclude: anthropomorphism, convincing characters and events, and generalizations of practical wisdom are essential components of the fable.

It is precisely these characteristics that set the fable apart from other similar literary models such as the fairy tale, the parable, the saying (proverb) or the didactic poem.

The fairy tale shares with the fable a common origin but magical elements make the difference. Speaking animals are part of the fairy tale, but the cast is normally larger, with a distinct preference for fabulous characters like dwarfs, fairies and giants. Moreover, the didactic purpose of the fairy tale is far less prominent and the happy endings guarantee that fairy tales are optimistic, while fables often have sad endings.

The fable differs from the parable through its clear symbolism. The fable can be understood immediately since it uses familiar, almost common knowledge. The strengths and weaknesses of the animals are well-known: the fox is cunning, the tortoise is slow. The specific allegorical nature of the parable, however, demands explanation. Swift's *A Tale of a Tub* (1704) and Lessing's Ring Parable in *Nathan der Weise* (1779) can only be fully appreciated if one understands that the objects (the coats and the rings) stand for different religions.

Both the saying and the fable are brief and to the point. What distinguishes the fable is the narrative character that underpins the message.

While the fable has a limited catalogue of figures (two or three animals), the didactic poem depicts in detail an image of society. Its message is more complex, philosophical and often politically engaged. *The Fable of the Bees* (Mandeville 1714) is, despite the word 'fable' in the title, rather a didactic poem.

In the fable, narrative record and lesson go hand in hand. A particular story is a metaphor for a general 'truth'. The meanings are versatile and have many applications.

Fables have often been used in a religious or moral context. In this case, fables typically proclaim that doing evil results in punishment, or that wealth causes worries.

Several scholars have stressed the social function of the fable. They consider that the fable originated from the need of working classes to express their bottled feelings and contest hierarchy and abuses of power (Hasubek 1982). The idea that Aesop, the first known fabulist, was a slave who acquired freedom and became the king's counselor strengthens this sociological interpretation.

In some fables the critical tone indicates a satirical approach. La Fontaine, in 'Les animaux malades de la peste' (The animals seized with the plague), provides a clear illustration of this. Some mighty animals unload the responsibility for a plague onto the naïve donkey who figures as the scapegoat. He confesses a peccadillo and is sentenced to death. La Fontaine is here criticizing abuses of power and the class justice of his times.

> What dreadful crime! eat other people's grass ! So for a hanging case they made it pass; Nothing but death could for the deed atone,Which to the ass was quickly shown.According as you're feeble, or have might, High courts condemn you to be black or white. (*Fables*, Book VII, 1)

Can the long tradition and the universal claims of the genre foster a sense of sustainability?

Can we trust the wisdom of fables and take them as personal guides to behavior, or general rules? Can the wisdom of fables still reach today's leaders or managers?

2 Fables About Sustainable Values and Economic Wealth

Almost all fables can be read on different levels. Morals typically address individuals but are often also appropriate in a larger (social) context, since the wisdom we honor in our personal lives is mostly also applicable on a professional level, in economic activities or in politics.

Several fables illustrate the Ancient Greek aphorism 'Know thyself'. Animals that deny their own nature and capacities are punished severely. They pay dearly with their lives, like the ambitious tortoise that wanted to fly, or the foolish seal that became a stranger among his own sort. Vices such as unbridled ambition, recklessness, overconfidence and pride cloud self-knowledge and bring on misfortune. Leaders in particular should take these messages into account, as 'The eagle and the arrow' teaches us.

The Eagle and the Arrow (Aesop—Perry 230)

An eagle sat perched on a lofty rock, keeping a sharp lookout for prey. A huntsman, concealed in a cleft of the mountain and on the watch for game, spied him there and shot an arrow at him. The shaft struck him full in the breast and pierced him through and through. As he lay in the agonies of death, he turned his eyes upon the arrow. 'Ah! cruel fate' he cried, 'that I should perish thus: but oh! Fate more cruel still, that the arrow which kills me should be winged with an eagle's feathers!'

In this fable the eagle does not take into account his own vulnerability. His agony is worse because he realizes that he will die from a weapon partly of his own making. The conclusion that we give our enemies the means for our own destruction hurts.

The Tortoise and the Eagle (Aesop—Perry 230)

A tortoise, discontented with his lowly life, and envious of the birds he saw disporting themselves in the air, begged an eagle to teach him to fly. The eagle protested that it was idle for him to try, as nature had not provided him with wings; but the tortoise pressed him with entreaties and promises of treasure, insisting that it could only be a question of learning the craft of the air. So at length the eagle consented to do the best he could for him, and picked him up in his talons. Soaring with him to a great height in the sky he then let him go, and the wretched tortoise fell headlong and was dashed to pieces on a rock.

The Seal who Became Famous (James Thurber)

A seal who lay basking on a large, smooth rock said to himself: all I ever do is swim. None of the other seals can swim any better than I can, he reflected, but, on the other hand, they can all swim just as well. The more he pondered the monotony and uniformity of his life, the more depressed he became. That night he swam away and joined a circus.

Within 2 years the seal become a great balancer. He could balance lamps, billiard cues, medicine balls, hassocks, taborets, dollar cigars, and anything else you gave him. When he read in a book a reference to the Great Seal of the United States, he thought it meant him. In the winter of his third year as a performer he went back to the large, smooth rock to visit his friends and family. He gave them Big Town stuff right away: the latest slang, liquor in a golden flask, zippers, a gardenia in his lapel. He balanced for them everything there was on the rock to balance, which wasn't much. When he had run through his repertory, he asked the other seals if they could do what he had done and they all said no. "O.K.", he said. 'Let's see you do something I can't do." Since the only thing they could do was swim, they all plunged off the rock into the sea. The circus seal plunged right after them, but he was so hampered by his smart city-clothes, including a pair of seventeen-dollar shoes, that he began to founder at once. Since he hadn't been swimming for 3 years, he had forgotten what to do with his flippers and tail, and he went down for the third time before the other seals could reach him. They gave him a simple but dignified funeral.

Moral Whom God has equipped with Flippers Should not Monkey Around With Zippers.

The importance of gaining personal self-knowledge is the primary message here, but the same holds for institutions and companies. They should be aware of their strong and weak points in order to make the right decisions.

Other themes are more directly focused on the public interest and the common good.

Sustainability concerns the worldwide community. It is linked to wisdom, frugality, providence and having long-term vision. But providence can also come into conflict with other values such as solidarity, as happens in 'The Grasshopper and the Ants'.

The Grasshopper and the Ants (Aesop—Perry 373)

One fine day in winter some ants were busy drying their store of corn, which had got rather damp during a long spell of rain. Presently, up came a grasshopper and begged them to spare her a few grains, 'For,' she said, 'I'm simply starving.' The ants stopped work for a moment, though this was against their principles. 'May we ask,' said they, 'what you were doing yourself all last summer? Why didn't you collect a store of food for the winter?' 'The fact is,' replied the grasshopper, 'I was busy singing, and I hadn't the time.' 'If you spent the summer singing,' replied the ants, 'you can't do better than spend the winter dancing.' And they chuckled and went on with their work.

In this well-known fable the ants apply themselves to their work and their diligence is rewarded. They foresee that the summer will come to an end and build up winter stocks. The singing cricket had a good time in summer, but will starve during the winter. The fable warns of the perils of imprudence and highlights that foresight is the essence of good government.

The text reveals to us that the ants are industrious and conscientious (even principled), but they are also impassive and even a bit heartless. They chuckle over the cricket's misfortune. Is providence always ambiguous? Could this fable put the European monetary crisis in a different light? Some observers consider the stubborn, hardworking Germans to represent the heartless ants, while the Greeks play the role of the singing grasshopper. Frugality and discipline versus spendthriftiness and a tendency to slack off. However, the fable also reveals some ambiguity: the ants lack generosity and compassion. There is no place here for solidarity.[4]

In another fable written by Aesop the hungry fox finds the food he needs, but he is so greedy that he causes himself harm.

The Fox with the Swollen Belly (Aesop—Perry 24)

A hungry fox found in a hollow tree a quantity of bread and meat, which some shepherds had placed there against their return. Delighted with his find he slipped through the narrow departure and greedily devoured it all. But when he tried to get out again, he found himself so swollen after his big meal that he could not squeeze through the hole, and fell to whining and groaning over his misfortune. Another fox, happening to pass that way, came and asked him what the matter was; and learning the state of the case, said; 'Well, my friend, I see nothing for it but for you to stay where you are till you shrink to your former size; you'll get out then easily enough.'

Only the fox himself is responsible for his misfortune: he maneuvers himself into this awkward and perhaps dangerous position. Just like the grasshopper, he doesn't think ahead and calculate the consequences of his actions. He acts on the spur of the moment in an impetuous way. Impulsive decisions and a focus on unbridled growth are also a constant threat in business and politics. The greed of banks and

[4] In recent times, the fable has again been put to political use by both sides in the social debate between those who support a culture of enterprise and those who consider that the advantaged have a responsibility towards the disadvantaged.

The Greek economist Yanis Varouvakis uses the fable in his blog of the 15th of December 2012. A modern satirical version of the story, originally written in 1994, has the grasshopper calling a press conference at the beginning of the winter to complain about socio-economic inequity, and being given the ant's house. This version was written by Pittsburgh talk show guru *Jim Quinn* as an attack on a social program of the Clinton administration in the USA. In 2008 Conservative columnist *Michelle Malkin* also updated the story to satirize the policies of 'Barack Cicada' (Wikipedia.org/wiki. The_Ant_and_the_Grasshopper#The_moral_debate) Retrieved 2012-06-22.

Alternative versions of this fable have been written by W. Somerset Maughan (*The Ant and the Grasshopper* 1924) and John Updike (*Brother Grasshopper* 1987). In both stories it is the merry 'grasshopper' brother who at the end is in luck, while the hardworking ants pass their days in solitude. In Leo Leonni's children's book *Frederick*, the poet who does not gather food is recognized as a valuable member of the group.

other financial institutions that pursued a policy of uncontrolled growth and created an unhealthy financial bubble resembles the greed of the fox with the swollen belly. 'Shrink to your former size' might also be the only wholesome remedy for them.

To look ahead at what is coming is also the main topic in 'The farmer and his sons'. But providence is a hidden element here that one just discovers on the way.

The Farmer and his Sons (Aesop—Perry 42)

A farmer, being at death's door, and desiring to impart to his sons *a secret* of much moment, called them round him and said, 'My sons, I am shortly about to die; I would have you know, therefore, that in my vineyard, there lies a hidden treasure. Dig and you will find it.' As soon as their father was dead, the sons took spade and fork and turned up the soil of the vineyard over and over again, in their search for the treasure which they supposed to lie buried there. They found none, however: but the vines, after so thorough a digging, produced a crop such as had never before been seen.[5]

The message that the dying father leaves to the next generation is that labor can bring reward. However, the message is coded as 'a secret' and the sons have to look for 'a hidden treasure'. What they get is not what they expected, but they are rewarded handsomely.[6]

The crops and the income they acquire for working is due to their own merit and may be a more sustainable good than an inheritance (money or gold) that is easily received but also quickly dissipated. Investing time and labor in the vineyard is the right thing to do create long-term benefit.

However, the happy ending is in a certain sense also an open ending. Did the sons really find what their father meant by his 'hidden' treasure'? Or was the rich crop of grapes only a stroke of luck?

To sum up, the lesson of this fable sounds very much like the Chinese saying: The way is the goal. Eastern spiritual practitioners have developed a set of theories around overcoming the duality of self and object and have tested and refined them through spiritual practice. The concept of flow fits here. Deep concentration on tasks may enhance workplace satisfaction and increase one's happiness. The fable advocates the search for meaning in work.

In 'The Miser and his gold' the reader is also warned that riches in abundance are not the main thing to strive for. The hoarding miser is unproductive and doesn't

[5] In Caxton's translation (1484) this moral is added: *For who trauaylleth wel / he hath euer brede ynough for to ete / And he that werketh not dyeth for honger*. In L'Estrange's Translation (1692) it is formulated as follows: *THE MORAL. Good Counsel is the best Legacy a Father can leave to a Child, and it is still the better, when it is so wrapped up, as to beget a Curiosity as well as an Inclination to follow it.*

[6] One could mention here 'the invisible hand' of Adam Smith. Benefits can sometimes be obtained even when they are unintended. In a more economic context: individual ambition benefits society, even if the ambitious have no benevolent intentions.

create any added value. His passive waiting is the cause of blame in the story and his loss of gold does not arouse pity.

The Miser and his Gold (Aesop—Perry 225)

A miser sold everything he had, and melted down his hoard of gold, which he buried secretly in a field. Everyday he went to look at it, and would sometimes spend hours gloating over his treasure. One of his men noticed his frequent visits to the spot, and one day watched him and discovered his secret. Waiting for his opportunity, he went one night and dug up the gold and stole it. Next day the miser visited the place as usual, and, finding his treasure gone, fell to tearing his hair and groaning over his loss. In this condition he was seen by one of his neighbors, who asked what his trouble was. The miser told him of his misfortune; but the other replied, 'Don't take it so much to heart, my friend; put a brick into the hole, and take a look at it every day: you won't be any worse off than before, for even when you had your gold it was of no earthly use to you.'

Echoes of the values that these fables proclaim can also be found in the parables of Jesus, such as the need for providence and alertness in the 'Parable of the Wise and the Foolish Virgins', and the need for effort and personal commitment in the 'Parable of the Talents or Minas'.

Several fables proclaim another message that is worthwhile hearing nowadays: *Festina lente* is a classical adage and oxymoron that means "make haste slowly", or "more haste, less speed". Suetonius writes in his *De vita Caesarum* that the emperor Augustus deplored rashness in a military commander. The motto 'festina lente' has been attributed to him.

He thought that nothing less became a well-trained leader than haste and rashness, and, accordingly, some favorite sayings of his were "More haste, less speed", "Better a safe commander than a bold", and "That is done quickly enough which is done well enough".[7]

'Slow down' is a frequently heard call, even in business. *Festina lente* is more complex. It rather means that activities (e.g. the way food is prepared according to the beliefs of the Slow Food movement) should be performed with a proper balance of urgency and diligence. If tasks are rushed, then mistakes are made and good long-term results are not achieved. This point is exactly the one made in the fable 'The Hare and the Tortoise'. La Fontaine formulates it like this: 'Rien ne sert de courir: il faut partir a point' ('It doesn't help to run, one must start in time').

The Hare and the Tortoise (Aesop—Perry 226)

A hare was one day making fun of a tortoise for being so slow upon his feet. 'Wait a bit,' said the tortoise; I'll run a race with you, and I'll wager that I win.' 'Oh, well,' replied the hare, who was much amused at the idea, 'let's try and see'; and it

[7] www.leidenunivnl/fsw/verduin/kerst/festlent.htm (2012.06.12).

was soon agreed that the fox should set a course for them, and be the judge. When the time came both started off together, but the hare was soon so far ahead that he thought he might as well have a rest: so down he lay and fell fast asleep. Meanwhile the tortoise kept plodding on, and in time reached the goal. At last the hare woke up with a start, and dashed on at his fastest, but only to find that the tortoise had already won the race. Slow and steady wins the race.

The tortoise here represents tenacity and stubbornness. Calmly she perseveres in what she started, but thoroughly and in her own way. The hare is overconfident, even reckless. He starts in a hurry but doesn't finish the job. The fable teaches us that steady work pays off, even when personal talents are modest. The fable indicates that perseverance can overcome being slow. Since the tortoise wins the race the opposition between slow and swift is overcome. The fable awards respect to moderation, equanimity and composure, attitudes that also suit the modern entrepreneur.

In order to have this long-term perspective one needs a dream that that can overcome the need for instant or immediate satisfaction. Spirituality can be a means to foster this dream.

The Moth and the Star

'The Moth and the Star', a fable written by James Thurber, helps us to get started dreaming.

A young and impressionable moth once set his heart on a certain star. He told his mother about this and she counseled him to set his heart on a bridge lamp instead. "Stars aren't the thing to hang round," she said; "lamps are the thing to hang around." "You get somewhere that way," said the moth's father. "You don't get anywhere chasing stars." But the moth would not heed the words of either parent. Every evening at dusk when the star came out, he would start flying toward it and every morning at dawn he would crawl back home worn out by his endeavor. One day his father said to him, "You haven't burned a wing in months, boy, and it looks to me as if you are never going to. All your brothers have been badly burned flying around street lamps and all your sisters have been terribly singed flying around house lamps. Come on, now, get out of here and get yourself scorched! A big strapping moth like you without a mark on him!".

The moth left his father's house, but he would not fly around street lamps and he would not fly around house lamps. He went right on trying to reach the star, which was four and one-third light years, or twenty-five trillion miles, away. The moth thought it was caught in the top branches of an elm. He never did reach the star, but he went right on trying, night after night, and when he was a very, very old moth he began to think that he really had reached the star and he went round saying so. This gave him a deep and lasting pleasure, and he lived to a great old age. His parents and his brothers and sisters had all been burned to death when they were quite young.

Moral Who flies afar from the sphere of our sorrow is here today and tomorrow.

This fable can be read on different levels. It can be seen as a love story, or on a more social or spiritual level as a story about social goals, a dream, passion or life fulfillment.

The moth cherishes a far-off dream, while his parents and siblings can satisfy their goals close at hand. But the immediate satisfaction of their aims literally causes 'burn out'. They all die young. The protagonist's long term vision may be unrealistic but it gives him energy and happiness and keeps him alive.

James Thurber inserts into his fables—like Aesop and La Fontaine—a sort of ambiguity. The hero of this fable reaches an advanced age and this in itself is a reward. On the other hand, it is probable that all those years he was chasing a shadow or an illusion, although a happy one. This is also the conclusion of empirical research that concludes that religion is a factor that increases happiness. Religion, spirituality or just wisdom is a means of self-transcendence.

3 Conclusion

The fable has proven itself to be a simple and successful narrative formula. It has stood firm in different cultures from antiquity onwards, and the criticism it has received from scholars has been predominantly constructive. Animal characters simplify the process of identification. Males and females, young and old, easily find a place in these stories. Characters are both flat and trenchant. The fable focuses in the first place on situations, traits and feelings. Characteristics are inflated to confirm the message that is often made explicit in a sentence or maxim. Often, several conclusions can be drawn. The lesson provokes an 'A-ha!' experience. The fable often offers us a new perspective for looking at the world. But, at the same time, the fable functions as a mirror that reflects the virtues and vices of its reader.

If we look at separate examples of fables the messages might at first sight appear to be simple. Looking closer, the characters and the situations are often more ambiguous. In the collection of fables as a whole, contrasting values tend to balance each other. This relates to the paradox of divergent problems, as formulated by Schumacher.

There is no simple answer to complex problems. The economic crisis exemplifies this fact clearly. Labor (the farmer and his sons), frugality (the ants), soberness (the fox) and productivity (the lesson of the miser) should come together if we are to overcome the crisis. The steady tenacity of the tortoise will help us to persevere—and don't forget the dream of the moth, which is an excellent guide to happiness.

References

1999. Aesop's Fables. (trans. by V.S. Vernon Jones). Hertfordshire: Wordsworth Editions.
Blackham, H. J. 1985. *The Fable as literature*. London: The Athlone Press.
De Vinci, L. 2002. *Maximes, Fables & Devinettes*. Paris: Arléa.
2001. *Dictionnaire des Genres et Notions Littéraires*. Paris: Albin Michel.

Hasubek, P., ed. 1982. *Die Fabel. Theorie, Geschichte und Rezeption einer Gattung*. Berlin: Erich Schmitt Verlag.

Jochum, K. P. 1982. Die Fabeln James Thurbers. In *Die Fabel. Theorie, Geschichte und Rezeption einer Gattung,* ed. Peter Hasubek, 236–252. Berlin: Erich Schmitt Verlag.

Jolles, A. 1968. *Einfache Formen: Legende; Sage; Myth; Rätsel; Spruch; Kasus; Memorabile; Märchen; Witz*. Tübingen: Niemeyer.

Liebfried, E. 1967. *Fabel*. Stuttgart: J.B. Metzlersche Verlagsbuchhandlung.

Pabst, W. 2002. Würde und Lob der Gelassenheit in Kunst und Fabel oder Die vier exemplarischen Karrieren der Schildkröte. In *Gestaltung—Ungestaltung*: *Beiträge zur Geschichte der romanischen Literaturen,* ed. B. König and J. Lietz, 265–86. Tübingen: Narr.

Thurber, J. 1989. The thurber carnival. New York: Harper & Row Publishers.

Rita Ghesquière is emeritus professor of comparative literature at the Catholic University of Leuven, Belgium. She holds a degree in philology and has written her PhD on Phenomenology and Literary Theory. Her research and publications fall within the scope of the history of European literature, juvenile fiction and spirituality. She published a Dutch literary companion on the history of European Literature. Together with Knut Ims she is the editor of "*Heroes and Anti-heroes. European Literature and the Ethics of Leadership*" (2010, Garant: Antwerp, Belgium, Apeldoorn, the Netherlands).

Part III
Innovative Practices and Policy Reforms for Sustainability

Sustainocracy: Spirituality and Sustainable Progress

Jean-Paul Close

Part 1 of this paper summarizes the author's perspective about human evolution, which includes observations about the quantum leap that we are about to experience that will be preceded by chaos and enlightenment. Part 2 describes the author's experience with a new sustainability venture in Eindhoven, The Netherlands, which takes the form of a new model for complex society moving toward sustainable progress.

1 A Quantum Leap in Human Evolution

To understand natural human progress in an increasingly complex human world, I have outlined a guiding model (Fig. 1) to make matters easier to understand. It consists of two intersecting lines. The first is a vertical line that represents *spirituality* (To Be). This refers to conscious inner reflection about the meaning of our existence and experience. The second, horizontal line represents the way we *act and organize* ourselves (To Do), using institutional structures, hierarchies, infrastructures, technologies, possessions, knowledge, etc.

The model provides insight and guidance for both individuals and the most complex human organizations. The historical evolution of any community can be drawn using these lines and quadrants that show the cyclical and spiral patterns of human development over time.

Our species has evolved around at least four dominant and fragmented spheres of interest and control: state (claims to territorial rights), religion (claims to spiritual truth), tools (knowledge and industrialized creativity) and valuables (material resources, possessions, money). Each has established its own dominance over people

J.-P. Close (✉)
STIR Foundation, AiREAS and STIR Academy, Eindhoven, The Netherlands
e-mail: jp@stadvanmorgen.com

L. Zsolnai (ed.), *The Spiritual Dimension of Business Ethics and Sustainability Management*, DOI 10.1007/978-3-319-11677-8_8

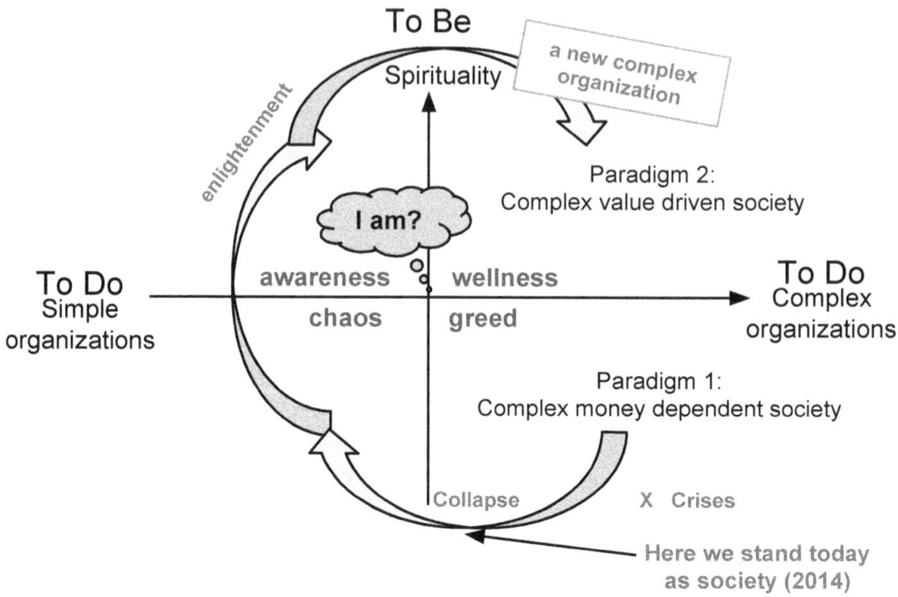

Fig. 1 Two paradigms of human complexities

through systems of dependence, lawful authority or indebtedness. Loyalty has been demanded in exchange for certain types of security or protection. We have learned how to manipulate the human mind and behavior using fear and the exercise of power over the masses. This type of fragmented dominance may have an evolutionary logic as a heritage of our animal origins. It has taken us a long time to learn how to deal with our natural but aggressive drive to survive, to compete, our aversion to pain or scarcity and our fear of death and even life itself. We have also needed to deal with our curiosity, our doubts and our awareness of the inexplicable miracle of our existence, awareness and surroundings.

Since our first spiritual quantum leap[1], we have developed the beliefs and the means to subdue our surroundings, nature, and other human beings to our will. This has caused a protracted battle with our own morality and with our conscience; a conscience that is gradually developing and goes hand in hand with our strong creative drive to understand and make sure things go as we wish (Fig. 2).

Crises have become natural events in this historical learning process of trial and error. The intensity of a crisis is directly related to the level of complexity of the society that collapses. The speed of recovery is equivalent to the level of spiritual reflection and the adaptive capacity of a community. Humanity learns about values, control and behavior the hard way. We eventually create periods of peace and wellbeing and then lose them again when we become too greedy, bureaucratic and selfish to sustain them.

[1] The first quantum leap occurred when we learned how to create and master fire. This led to humans' lasting sense of divine superiority over their surroundings.

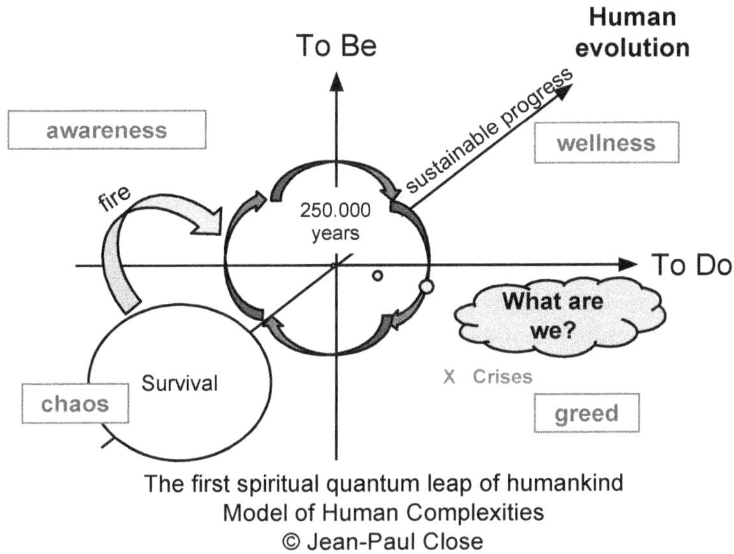

Fig. 2 Human evolution

We now realize that anticipation, reflection, adaptability and teamwork are necessary for the process of voluntary change towards value-driven progress. We unite personal knowledge, scientific understanding and technological means, and trust our ability to produce abundance for each human being under the most adverse circumstances. However, we lack the individual and collective spiritual awareness to treat humankind and sustainable progress as a whole. The opportunity to create social and cultural cohesion and work as a global team has only become a realistic option over the last few decades. Our common purpose has recently been placed on the global agenda as an intention waiting for a breakthrough. This breakthrough—humanity's second quantum leap—is happening now.

Ultimately, spirituality can reach out to foster the individual's deep inner awakening (I am) to a higher level of consciousness and meaningful personal responsibility (To Do) within the context of our evolutionary reality as a species (Sustainable Progress).

For many millennia human societies and civilizations went through a process of experimental cyclical evolution. About 3000 years ago the population had become so large and so advanced in their ability to travel and use technology that encounters between civilizations were frequent. Interaction necessarily became more diplomatic and demanded ways of exchanging knowledge, resources and technology and thereby inspired humans to organize societies in a peaceful yet productive way.

Human development then accelerated exponentially, with increases in awareness shortening each cycle from millennia to centuries to the decades it takes today. With a life expectancy of only 30 to 40 years, few of our ancestors had the opportunity to accumulate enough personal experience to reach a higher level of consciousness.

Since then, technological and social innovation has increased exponentially (particularly over the last 300 years) as industrialization and warfare further enhanced humankind's knowledge, awareness and interaction. World War II, the last major confrontation of global proportions, scared humanity to its core, producing a huge cycle of crisis, chaos and global enlightenment everywhere. Humans have become extremely aware of their dark side and the tremendous destructive potential of their technologies. The organized processes of mass murder that occurred in the past have now become shameful and well-documented modern precedents that can no longer be manipulated by politicians, neglected by the public or romanticized by historians. It is in everyone's interest to maintain peace and avoid war. Technology should support progress, not destruction.

The old fragmented hierarchies of power have now been shaken up due to the channeling of interests into money-based structures. Since World War II the world has been undergoing a historic process whereby global human interests have tended towards unification. Yet the many fragmented and resilient hierarchies that still exist, based entirely on external forms of security, have no spirituality. Their meaning is defined by selfish desires for power, control or individual self-enrichment, illustrated by their repeated call for (economic) growth. The monetary system has caused isolated expressions of power to be grouped into a unified system so that for the first time in history the entire human community is focused on the collective, highly-organized and 'well-educated' goal of resource acquisition. In the modern system of financial debt and greed we are all "emancipated" and equal.

Today, the average, well-educated modern human being with a life expectancy of 80 to 90 years and instant access to universal knowledge goes through several cycles of awareness in a single lifetime! Such an individual has every opportunity to break through to higher personal spiritual consciousness and still have plenty of time left to put this learning into practice. Never before in human history, after thousands of years of evolution, has such a unique combination of individual awareness and collective crisis come together. Yet it stops here!

After examining the situation of humanity at the present time as a whole it is possible to draw the situation using a model of human complexities (illustration below) for the start of the twenty-first century. A singularity has been reached.[2] It is easy to be filled with fear when we observe the world collapsing because our sense of security is taken away with the collapse. We may remain blind, hoping that a new world will be created through a paradigm shift. But this human world already exists. It is just a matter of becoming aware of it and making it happen by taking on personal responsibility. We can lose all we "have". It's what we "are" that protects us in our progression (Fig. 3).

The beginning of the twenty-first century will enter the history books as a brief but intensely significant period in human history when humankind finally became painfully and powerfully aware of its spiritual self. This is the second quantum leap in human evolution[3] and will result in a new, self-aware global human society that

[2] A point of singularity is a point at which an exponential development cannot reach infinity and therefore collapses.

[3] The second quantum leap refers to the conscious organization of our complex global society around our spiritual enlightenment, making higher consciousness a leading criterion for creating sustainable progress. The first leap taught us how to use external fire. The second leap will show us how to use our inner, spiritual fire.

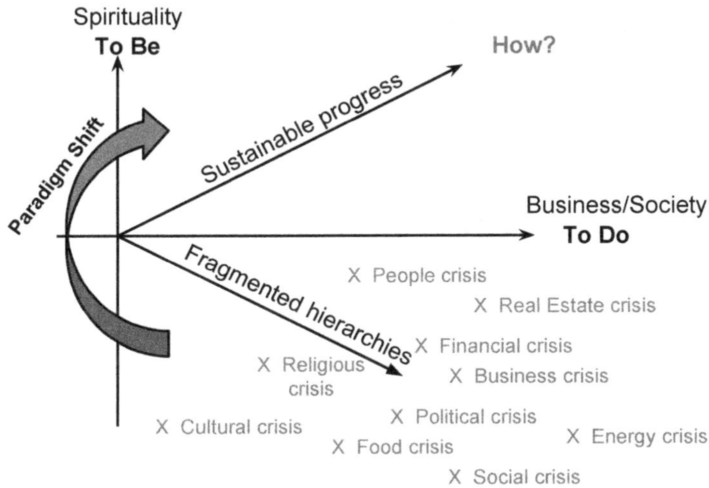

Fig. 3 The end of 250,000 years of system dominance, spiritual confusion and negligence

remains after the singularity. Instead of 'having' (possession) and 'taking' (greed) we are learning to 'be' (talent) and to 'give' (create). We are learning to be before we do (Fig. 4).

2 A Practical Transition and Working Model for a New Society Based on Sustainable Progress

With the majority of members of modern society focused on organized greed, one of the first steps in creating the new model was to define "sustainable human progress" to create a point of reference and common understanding for the new paradigm. The

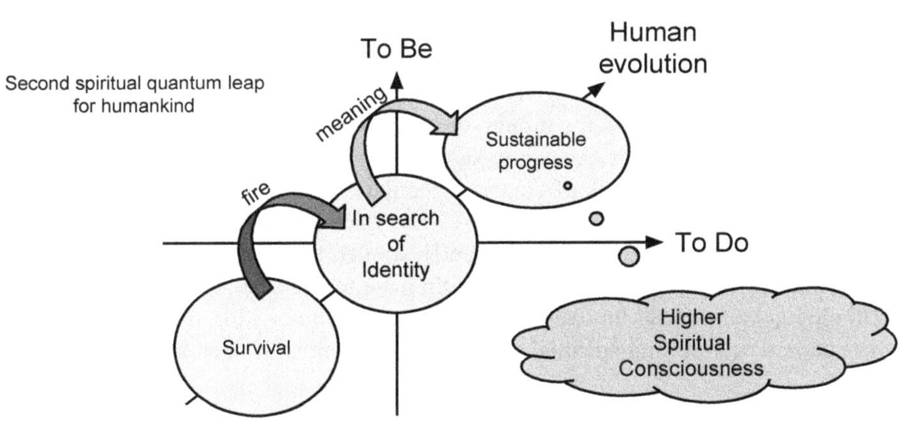

Fig. 4 The second quantum leap occurring now

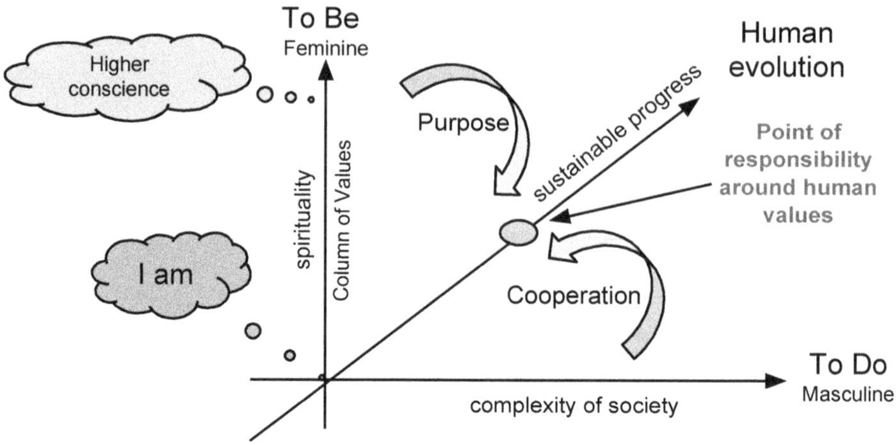

Fig. 5 Sustainocracy: the new model for society

1987 Brundtland definition used by the United Nations is useless for this purpose as it defines sustainable development from within the old paradigm, based on greed.[4] This does not work, as I experienced in my quest to create security for my family. I needed to define something which I (and everyone else) could take direct responsibility for—personally and institutionally, right now and all the time.

Thus my definition of sustainable progress is as follows: *to continuously work together in a healthy, vital, safe and dynamically progressive, self-sufficient human society within the context of the ever-changing natural environment in which we live and act.*

Still living within the old paradigm, I realized that we still need money for all our basic and non-basic needs. All the unique benefits that human evolution has brought us are available to the highest bidders. By remaining within a money-driven structure I would be vulnerable to speculation and manipulation, even for my basic needs. I also realized that life today is so complex that self-determination and self-sufficiency require personal awareness but also intense cooperation between many like-minded and talented individuals (Fig. 5).

By locating myself in the new value-driven paradigm (on the green dot of evolutionary responsibility) with the goal of ensuring a safe environment for my children, I would still have to access the money-based system if we wanted to eat, drink or have a roof over our heads. Access to such things is dependent on money, with its related and inherent systems of control, speculation and manipulation and subsequent crises. In the existing paradigm we are 99% dependent on others through money. If my own country, the Netherlands, wanted to be totally self-sufficient, it would need to be 20 storey high or it would have to use scientific know-how with technological and social innovation to create abundance within the confines of its own space. The resulting spiritual shift will be enormous if such a choice towards

[4] The Brundtland report, "Our Common Future" (1987, Oxford University Press, Oxford).

independence is made. If not, we will grow increasingly vulnerable in competition with the rest of the world simply because we have taken from others (including nature) to sustain wealth that is not ours.

Every institution in the old chain of relationships should be challenged to transform to fit the new paradigm and to use responsibility with talent differently. I believe this is a necessary change. Yet, while voluntarily letting go of the old paradigm may overcome the eminent threat of all the forthcoming painful crises that will be of unprecedented magnitude, it will not solve the problem of the vast emptiness of new, self-sufficient, local societies that will still need to find ways of meeting even basic needs. Entering the new paradigm, we will start creating society from scratch. After recognizing this challenge it became important for me to ask others to take responsibility too; not just self-aware individuals but institutions as well.

For a long time I was very much alone in my sense of urgency as most institutions around me had a blind focus on material survival, growth and self-interest. I organized and participated in awareness congresses and seminars, wrote books and articles, set up a foundation, offered consultancy services and attracted thousands of people to my views. But I did not manage to break through. If a thing did not generate money quickly, people did not consider it worth the effort or investment. Wherever I looked, money was still more important than value, even though we were polluting our environment and quality of life at an incredible rate.

In the end, all my initiatives led me to address an important issue that was not 'economized', nor taxed, yet was a serious enough problem in my own area: *air quality*. I decided to stop asking everyone else to bear some of the responsibility and decided to take responsibility myself. My main objective became to create a precedent through a problem-solving model, despite the opposition of the old world of bureaucratic power.

Air pollution has been a major issue for decades around the world and especially in the Netherlands. The economic growth of central Western Europe within the context of the money-driven paradigm has densely populated this region with activities that are highly polluting. This has been good for the economy but bad for our health. Statistics show that thousands of people die prematurely in the Netherlands each year of heart and lung problems directly related to air pollution. As a result, government expenditure on healthcare has grown tremendously over the years, responding to the issue but not solving it. The difference between both paradigms became clear. You can not stop a consequence-driven economy on which the entire region depends because of invisible matters like air quality and human health alone. The current money-driven chain of authority has limited instruments for dealing with the problem. It became clear that money could not solve this problem; more was needed. A complex moral and structural transformation was required.

The local complexity around air quality and human health provides a direct link to my definition of sustainable progress. It became clear to me that one needs to address the four key areas of authority that exist within any society if complexity is going to be tackled. In my case, these focal areas were:

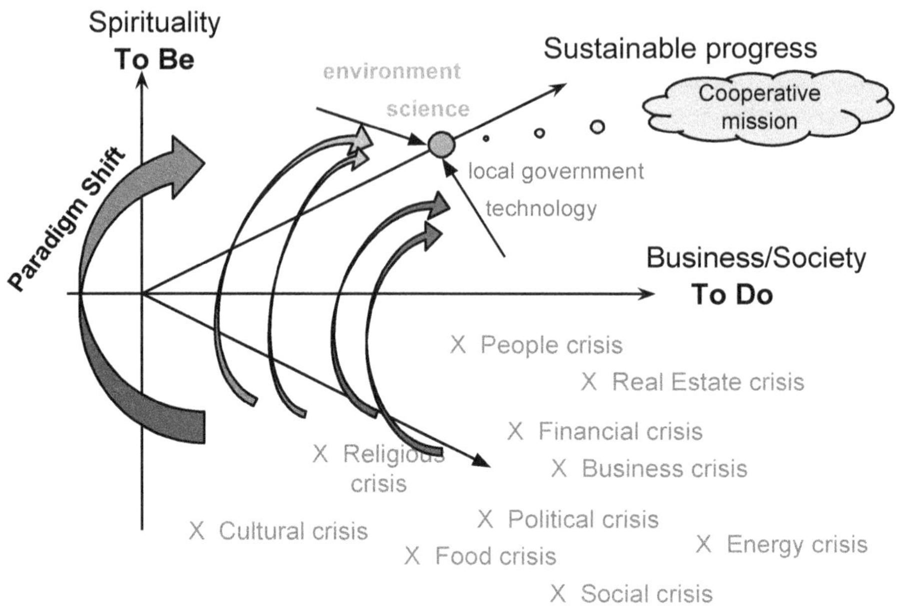

Fig. 6 AiREAS in the field of wellness

1. *The local population*: All the people living, working and visiting the city of Eindhoven are directly responsible for their share of environmental problems within the context of local circumstances (climate, soil, geographical location, culture, industries, life style, business, etc.) and so are also part of any solution,
2. *The local government*: The regional institution with its environmental authority that facilitates the progress of the local community using public means,
3. *Business enterprises*: Local and multinational entrepreneurial operations that supply creativity, technology and innovation,
4. *Science and education*: Providing scientific knowledge about air pollution, climate, the effects of different types of pollution on people, etc.

None of these community authorities normally relate to one other in a purpose-driven way, nor take responsibility for such complex co-creative problems. They all function as part of a chain of economic dependencies and opportunistic money-driven self-interests, each playing some fragmented, consequence-driven, incomplete role in the complex puzzle. Nevertheless, each of the four key fields of authority has a particular (often opportunistic but highly differentiated) interest in participating in a solution-oriented venture.

Instead of asking them all to take individual responsibility I decided to invite them all to take responsibility with me *together* in a venture with its own name: Local AiREAS Eindhoven (Fig. 6).

The financial and relational inter-dependencies of institutions in the reigning paradigm create new challenges. Who takes the lead in such a purpose-driven, new-paradigm venture? Each of the four authorities we needed to cooperate with looked at the others from the perspective of the old paradigm; as clients and suppliers in an economic chain of relationships, not as partners in a shared humanitarian mission. And the local population had no sense of responsibility at all. Their perception of consequence-related issues was that government should take care of everything.

I realized that if I were to decide to take responsibility by representing the new paradigm and the purpose of the AiREAS venture through promoting the line of sustainable progress, the old chain of dependencies could be broken. Together with a well-connected partner in the world of technology, Marco van Lochem, I started AiREAS as a co-creative mission in which all four areas of responsibility could be represented equally. When we invited the institutional authorities to join AiREAS they needed to break out of their chain of dependencies. It became their own institutional choice to participate, whatever their motivation for doing so, in a spirit of co-responsibility towards the mission, not to the fragmented interests that rule the old world.

Participants had to get used to the formula that I started using to guide the process:

Talent (institutional competences) × *Energy (motivation and authority)* × *Purpose (AiREAS)* = *Sustainable Progress (Measurable Human values)* × *Reciprocity.*

"Talent" was easily understood and translated to the daily reality of the participants. The multiplication of talents and authorities within the multi-disciplinary context of AiREAS even created a direct boost to focused professional creativity. "Sustainable Progress" could be explained in terms of financial, humanitarian and ecological benefits, and each of these benefits spoke for itself.

The other two, "Energy" (what motivates me to participate and provide input?) and "Reciprocity" (what's in it for me?) were more complicated. They were sources of major internal battles that required that the concepts were defined and agreed on by each of the disciplines in their own fields of interest.

For instance, local government officials had to come to terms with the new position they were in concerning their old monopoly on public executive authority, which also controls public taxpayers' money. For many centuries government has been responsible for spending taxes solely according to its own criteria, involving huge amounts of bureaucracy, internal teams of specialists and generalists, external consultancies and interaction with commercial lobbies. This world was turned upside down when a single individual invited them to join Local AiREAS. The entire bureaucratic apparatus was instantly eliminated from the process and the hierarchical power of local government was repositioned into territorial responsibilities. Officials had to interact with others based on the principle of equality and learn how to facilitate talent and create the means for local wellness without prejudice or executive monopoly.[5] This was accepted by operational officials but not by the bureaucrats involved. The entire underlying system of legal government participation in

[5] Equality is a third fundamental requirement for building trust in purpose-driven forms of cooperation (in addition to institutional identity (I am) and safety (respect, authenticity and integrity), according to the column of spiritual values that I created and use to guide these ventures forward.

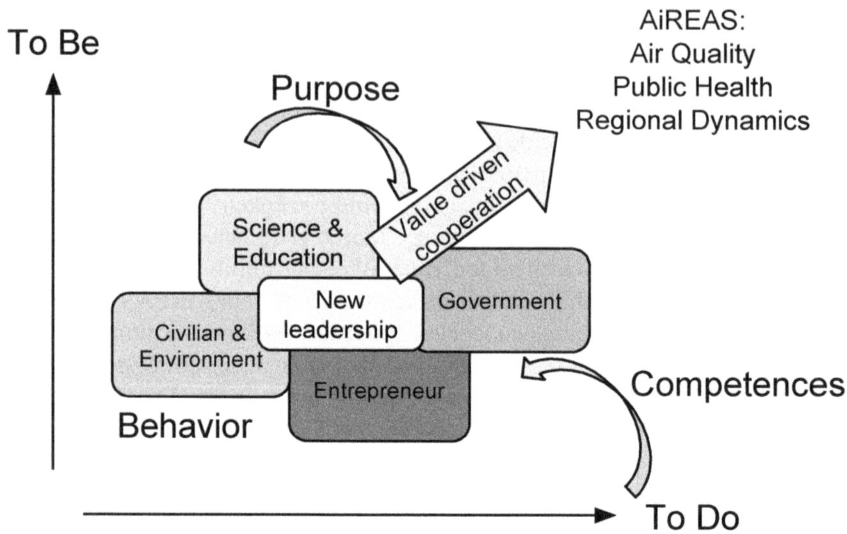

Fig. 7 Create a healthy and vital city using air quality as a point of reference

purpose-driven ventures (where they also act as financial partners and auditors) was also now questioned. Laws have been created around liability, controls and regulation, not around shared responsibilities, co-creation and responsibility for progress. The current structures of law were identified to be blocking sustainable progress. The moral question arose: what justification does a legal system (in which all the authorities of a free, democratic country are represented) have to block sustainable progress? On the one hand, there exists a democratically elected government that maintains the old system with its executive officials that address political concerns. On the other hand, the need has arisen to change constitutional bureaucratic structures to support models of sustainable progress such as AiREAS (Fig. 7). This is why sustainocracy was introduced as a new model for a complex society:

Sustainocracy = *Sustainable progress (no choice possible)*
× *democratic prioritization*

The Value-Driven Purpose of AiREAS

This instant change in the positioning of government responsibilities had another important effect. All the other parties had to adjust their behavior too! They had to understand their position from a co-creation point of view, accepting their own responsibilities towards the Local AiREAS Eindhoven mission instead of trying to sell their ideas to the government or consumers. All the participants had to work together and become part of the AiREAS team, participating in a result-driven venture that was measured through how well it met humanitarian and environmental, not by financial, progressive values.

The effect on the participants was huge—personally, as professionals contributing to a higher cause, and for each institutional identity. The latter needed to consider its motivation for participating by looking at its own inner commitment to this purpose-driven paradigm. Spirituality took the lead thanks to the meaningful purpose of the venture.

Many individual professionals considered participating in the all-inclusive venture but disappeared when they could not find the short-term economic self-fulfillment of the old paradigm. Others asked themselves what the consequences would be if they did not participate. The global transformative shift became clearly visible, as did the vulnerability of those who remained in the old paradigm. A new question arises for those who joined the new paradigm: what institutional reciprocity is possible in a world where the goal is human progress instead of money? This question opened up a totally new field of creativity and awareness for determining entrepreneurial value and return.

Local AiREAS Eindhoven quickly and organically auto-selected value-driven partners from opportunistic hunters by unconsciously addressing their "to be" awareness. This did not mean that the remaining partners were not still driven by their interest in money. Being part of the AiREAS venture offered competitive and strategic advantage too in terms of the opportunity to become a global pioneer in sustainable progress. The transparency of attitudes and shared responsibilities created a passionate community of purpose-driven professionals willing to go further than was institutionally expected of them. They started challenging their own organizations to provide the means of change, and some even quit their jobs when they felt they were not receiving institutional support. Others sought solutions by involving yet other institutions and departments that could help the venture proceed and overcome obstacles. Purpose-driven teamwork, technology, trust in one other, adaptability and a problem-solving attitude became more valuable than money alone. We had something to prove, first to ourselves and then to the rest of the world.

Within 4 months, various projects of great complexity were defined with the direct involvement of all four areas of competence. All the participants looked for financial and technical resources, not just the local government. Everyone took equal responsibility, taking risks along the way and sharing their successes and failures in all openness and transparency. The group became adaptive, focused on taking practical steps, and adjusted the steps to on-the-ground reality as it presented itself in the dynamics of complex, co-creating relationships. The only burden remained the lawful bureaucracy of an obsolete complex system.

Each of the participants (individual professionals as well as the institutions they represented), went through the following mind shift when they decided to join and remain in the Local AiREAS initiative.

- *Understanding the different paradigms*
  When entering the domain of Local AiREAS each participant was invited to take professional responsibility in the venture, not as a client or supplier to another but by providing their own unique key competences and authorities for sustainable progress and its overall results, as desired for human health and vitality. As a result they glimpsed the two worldviews.

- *Transformative leadership*
  Once the different worldviews were consciously or unconsciously understood, each of the participants needed to make a personal and institutional choice to be part of the venture or to leave it. Such a decision is intense as it represents a commitment to a higher purpose that can be underpinned by personal commitment, but not necessarily seconded by the participating institution that one represents. Members are challenged to commit their personal identities and asked to reflect deeply on meaning at an individual and institutional level.
  If one decides to continue in AiREAS a process of transformative leadership commences wherein the values to which participants commit should deliver a return that differs from that offered by a money-driven community. Personal or institutional satisfaction and continuity produces a conflict with the old world and requires a new positioning of participating identities. The larger the organization, the more difficult such a commitment for transformative change is. Small, local enterprises tended to be far more inclined to commit their talents than self-employed professionals (very much in survival mode due to financial dependencies) or large multi-national institutions (extremely dependent on global financials and volumes rather than on creating local value). Nevertheless, large institutions have the power to leverage local transformative change in far greater (and more complex) geographical or technological proportions.
- *Adapting competences*
  In a multidisciplinary venture, the expertise that unites is directly related to the type of result-driven projects that are defined by the coalition partners. When new disciplines are included, new insights, constructive fresh ideas and result-driven propositions emerge. Participants challenge one another to excel through using their own expertise, not by presenting products but by showing the effects of their contributions to the desired result. It was up to the participants to define their own criteria for success by investing interest and creativity in the venture rather than sitting back and waiting for random opportunities. The multidisciplinary mix of people and institutions produced a delicate blend of abstract idealism and pragmatic realism, often strongly guided by the direct presence of scientific know-how.
- *Cooperation*
  Multidisciplinary cooperation produces challenging results which one can reflect on to create further action. Local AiREAS Eindhoven had a direct effect on the entire process of local policymaking. Multidimensional links are being made by combining open source information with cross-relational research and communication. This is resulting in new policies for government officials and ideas for product innovations for entrepreneurial executives and business pioneers, as well as new educational and research opportunities for the universities, high schools and scientific institutions that are participating.

Reciprocity refers to the return that a participant receives from a venture like AiREAS. It refers to something more diverse than simple financial benefit. Each of the institutions gains spiritually from its contribution to the venture, strengthening their prospects in both the short and long term due to their public commitment to sustainable progress. But value needs to revert back to increasing the continuity of the organizations as well. At first, the Local AiREAS Eindhoven community was partly accepted by local governance out of fear of penalties imposed by the European Union in Brussels if certain normatives were not met by 2015. A few months later, the same community had developed sufficient insight and internal power to potentially challenge obligations set by the European Parliament by referring to the new standards that had been cooperatively established in Eindhoven to measure progress. This is the type of strength that such purpose-driven ventures can develop. The world-wide perspective of this innovative venture was considered interesting enough for highly competent, self-aware technology-focused multinationals to get involved, expecting to position themselves globally not just through their products but also the responsibility they profess to humankind and to progress.

None of the institutional organizations had any experience with fostering civilian, purpose-driven participation, even though the entire venture had been a civilian initiative from the start. Most of the institutional professionals who were involved live in Eindhoven and are, of course, civilians at work and at home. Why should the values defended in the context of institutional responsibility be any different from those promoted at home?

Many AiREAS participants have children that go to schools that may be health hazards because of their proximity to dense traffic arteries, or have relatives or acquaintances with lung or heart problems. When taking participative responsibility for sustainable progress, professional and private spheres merge. Involving 220,000 local people of a great diversity of ages, activities, motivations, occupations, etc. became a major issue that was encountered with apprehension and reservation. Old-paradigm institutions are used to being authoritative towards the public through regulations, laws, one-way marketing communication and financial debt constructions. The Local AiREAS initiative turned this relationship upside down. The general public was invited to take personal responsibility for its own environment and progress. This became a unique AiREAS challenge for which all kinds of techniques for public involvement and behavioral sciences could be tested, deployed, reflected upon and documented.

Allowing a population to come up with its own ideas and activities and to take responsibility without any institutional steering (only providing open source information) is a challenge. It is very difficult for dominant institutions to have faith—to let go and allow things to happen—without prejudice, interference or guidance. This requires having trust in the creative abilities of the highly-educated, developed environment and in the sense of responsibility that people can develop if they are given a chance, need and purpose of their own.

Local AiREAS Eindhoven was split into two professional teams; one for the hard innovation side of things (such as technological infrastructures, measurable information and research), and one for the soft side (represented by communication and listening techniques, public awareness and participation, behavioral changes and building non-judgmental trust in civilian initiatives). The soft side was placed in the able hands of local entrepreneurs, often self-employed professionals and small, creative and highly-enlightened offices instead of the more control-oriented institutional partners at the hard end. The same format for involving actors in the new society was respected but the multi-disciplinary players around the table changed.

High levels of open source communication and good examples of behavior (including that of institutional officials and structures) were made visible to the general public. Public recognition of good practice became key to developing a "can do" local civil culture.

Numerous spontaneous, unpredictable and highly innovative spin-offs emerged from the Local AiREAS Eindhoven movement. Social participation and involvement was followed up by professional AiREAS teams that translated data about air quality into instant feedback for the public. Provision of real-time information provided direct input for infrastructural changes and public support, making government a proactive facilitator rather than reactive regulator. The image of the government as a partner to the general public was strengthened.

The combination of hard institutional cooperation and soft public involvement in genuine purpose-driven Local AiREAS processes are expected to produce quick results—less than 12 months after the very first encounter, including all the complex spiritual awareness and transformational processes. Short-term successes are keeping the interactive coalitions together. We expect to see true long-term impacts in terms of an increase in environmental quality within years rather than several decades and this will benefit the overall health and vitality of the community. This, in turn, will improve regional dynamics, social cohesion, the overall commitment to sustainable progress and local value-driven productivity. Financially, the entire AiREAS mission cost just a minute fraction of the amount that the government had invested in previous years. Meanwhile, the changes are on-going—exponential, multilevel and lasting.

The same sustainocratic model, used for understanding and dealing with the societal complexities around sustainable progress, can instantly solve any crisis produced by the old paradigm. Value-driven purpose that unites talent and energy with local forms of reciprocity needs to be holistic, abstract enough to break through attitudes of fragmented individualism and pragmatic enough to create result-driven cohesion. But once the purpose of such ventures has been defined, the matter of whether to participate is still up to the individual or institution. In the near future, whether people and institutions decide to take responsibility when invited to do so will be an important indicator of their moral stance. In other words, the paradigm shift is not a process; it is still a choice and as such can be made instantly, anywhere in the world. As the crises deepen around the world, pursuing sustainocracy will become an act of institutional responsibility that can be evoked by entire populations. Higher consciousness demands that pioneers like me stand up and anticipate this development by inviting institutional authorities to take local responsibility now.

When local-for-local, purpose-driven communities define their visions of new local society they can all develop their own identities and join a purpose and result driven community of higher territorial abstraction without the intention of

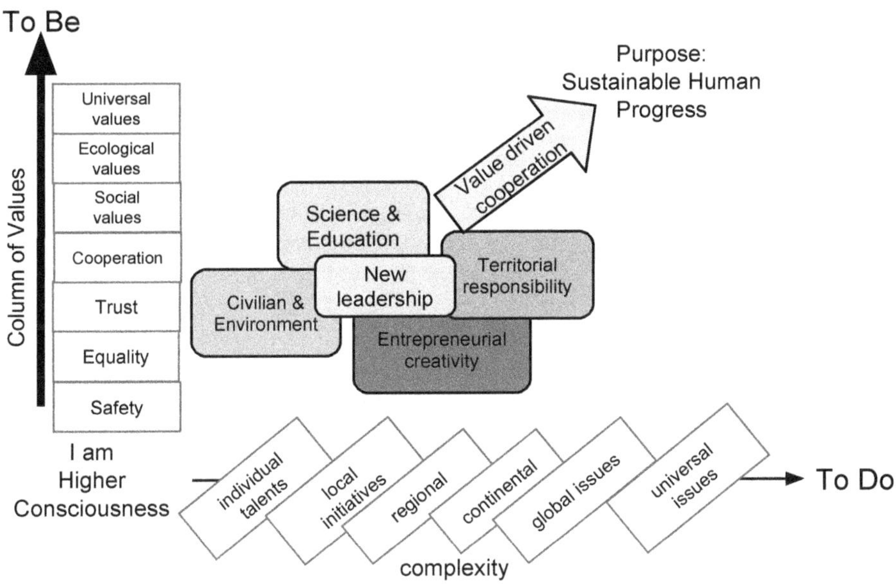

Fig. 8 The new local and global society

re-creating hierarchical structures. Local-for-Global then becomes a purpose-driven fact driven by sustainable progress until the next quantum leap occurs somewhere in the future (Fig. 8).

To help people understand the practicalities of the transformation we have created a game. The game is played between students, city officials, businesses and visionary coaches. They are asked to visualize their own city of tomorrow and subsequently work together to make it happen. The game is first played for a single day in school or university and later for a year by students who develop their abstract views into practical propositions. It is a safe way of getting acquainted with the model without challenging the old paradigm of the "real" world.

3 Conclusion

The Local AiREAS Eindhoven experience shows that a new format for society can be quickly developed and implemented to foster true sustainable progress in complex societies using powerful instruments from the old paradigm. This process represents an act of responsibility that challenges old bureaucratic dependencies and structures in a proactive and powerful way. Through voluntarily testing the new society using the stress points of human complexity as it increasingly appears over the coming years, the suffering of major fractions of the human populations may be reduced to the minimum needed for evolution, converting fear into pro-active passion and contribution.

In Eindhoven I was lucky enough to find a unique combination of the right people with the right mentality for change, in the right positions of authority at the right time. They trusted me when I asked them to come and take responsibility together for just one area of complexity. As we have just started, a lot of work still needs to be done. Yet the desire is there. If we are able to establish such local collaborations globally, we will eventually produce sufficient proof of abundance for everyone on our Earth. We may even learn to act like a global family, ready to take on our next evolutionary challenge whatever that may be—maybe even to populate the universe. The quantum leap will then occur and will be a memorable event in our history books—the moment humankind took responsibility for its own sustainable progress, with lasting spiritual love, not fear.

Jean-Paul Close After an executive career across the world Jean-Paul Close came back to his homeland, Holland in 2001. As an entrepreneur he started to write books, speak in public and coach enterprises on business spirituality. In 2008 he started to uniting entrepreneurial people with a high sense of cultural creativity in the STIR Foundation with the objective to work together on transformative processes. Starting with self-employed people they caught the attention of larger organizations and institutions to eventually create new cooperative initiatives in various fields. Proving their concepts slowly in the complexity of the Netherlands they started communicating their findings to international audience and through international papers.

Quakers and Climate Change

Laurie Michaelis

1 Introduction

The European Council has agreed that industrialized countries need to reduce greenhouse gas emissions (GHG) by 80–95 % by 2050, in order to limit global warming to below 2 °C (CE 2009). This goal looks increasingly unattainable. Achieving it would require 5–8 % annual cuts over the coming 40 years. A large part of such cuts would have to come from reductions in the level of material production and consumption—i.e. in the material standard of living (Michaelis 1998). At present, governments and the general public seem unable to engage with this prospect.

There are many diagnoses for this impasse. There is the problem of knowledge, truth and 'denial': if only people acknowledged what is happening, they would act (Weintrobe et al. 2013). Then there is the problem of power: the role of the fossil fuel industry and producer nations in protecting the status quo, and the need to empower people to change their lives (Berners-Lee and Clark 2013). And there is the problem of love and caring, and the limits on what and whom we include in our sphere of compassion, thinking that if only people had a relationship with nature they would act to preserve it.

Sustainability discussions in industrial societies are framed by a variety of cultural approaches to power, truth and compassion (or who is included in the sphere of compassion). Based on Cultural Theory (Douglas and Ney 1998; Thompson and Rayner 1998) some of the dominant cultural forms can be characterized as:

1. Hierarchies in which there are clear rules and lines of control. Truth and knowledge are handed down, whether from tradition or from rulers, experts or priests. People are cared for provided they can fit into an established role in the system.
2. Systems in which individuals or groups compete for success. Truth is important insofar as it influences success but people may believe what they like. Knowledge is private property that may be traded, created through innovation, or

L. Michaelis (✉)
Religious Society of Friends (Quakers), Oxford, UK
e-mail: laurie@livingwitness.org.uk

L. Zsolnai (ed.), *The Spiritual Dimension of Business Ethics and Sustainability Management,* DOI 10.1007/978-3-319-11677-8_9

obtained through experience. People and teams or firms are included until they are eliminated in competition.
3. Egalitarian groupings which emphasize dialogue, co-operation and mutual care. Truth and knowledge are negotiated and adopted through discourse. These groups normally have strong boundaries based on kinship, long academic or professional training, or religious initiation.

Much of politics and social change in Western democracies has been about shifting the emphasis among these three forms. However, they usually coexist in any one institution or group of people. For instance, to function well, markets and competitive sports require clear rules and dispute arbitration (hierarchy). The scientific community working on climate change shows signs of all three forms: there are egalitarian or collegiate dialogues among specialists, who also compete both together and against each other for research funding, and are organized within the hierarchies of their research institutions and professional bodies.

Climate change can be seen as a problem arising out of an overemphasis, both in government policy and in popular culture, on free market capitalism and competitive individualism. This system has supported innovation to create technologies and products that have led to unprecedented improvements in the standard of living in industrialized societies. It has also enabled a concentration of wealth which is often linked to political power, or to position in the social hierarchy. Hence it has created considerable vested interests in maintaining the status quo.

The culture of competitive individualism is not well-suited in itself to responding to climate change:

1. its relativistic approach to truth means that authoritative findings from the scientific community are easily disputed and dismissed as 'beliefs'
2. its commitment to liberty in the pursuit of individual self-interest makes it hard for people to consider any limitation on consumption
3. it tends to view both nature and humanity as resources to be developed and used in pursuit of economic success.

Many sociologists and economic historians (including Rostow 1990; Schumpeter 1943) have followed Marx in anticipating that free market capitalism will eventually be replaced by a more communitarian, egalitarian system. Beck (1988) sees signs of a process towards a more discursive culture in what he terms 'reflexive modernization': industrial society becomes increasingly reflexive in that its external impacts rebound upon it; interest groups proliferate, criticizing powerful institutions. This leads to increasing public reflection on and discussion of the policies, decisions and assumptions that shape organizations and society as a whole. Institutions become 'post-traditional' (Giddens 1994). The process has been facilitated by new media, which turn mass broadcast audiences into participants in discourses.

While interest groups on climate change may be part of a shift towards a more egalitarian, discursive way of running society, many devote their energy to criticism of government policies and corporate strategies, adding to the confusion of voices and the sense that climate change is a contested issue. However, several community- and group- based initiatives are trying to demonstrate the potential for constructive and inclusive dialogue, practical action and lifestyle change. These include the Global Action Plan EcoTeams project (Burgess 2003), Carbon Reduction Action Groups (Howell 2012) and the Transition Movement (Hopkins 2013).

In 2001, I started *Living Witness Project* with the aim of supporting a number of Quaker groups to experiment with collaborative approaches to sustainable living. In this paper I hope to share some of the experience and insights arising from that work, and from the distinctive Quaker approach to power, truth and inclusion.

2 Why Quakers?

Quakers internationally vary in their approach to religious tradition and the extent to which they consider themselves Christian or take the Christian Bible as an authority. In talking about 'Quakers' or 'Friends' in this chapter I will refer primarily to Quakers in Britain and other 'liberal' Quaker traditions.

Quakers represent a particular blend of egalitarian, hierarchical and individualist values and structures (Michaelis 2008). At first sight the Society of Friends epitomizes many of the developments in egalitarian and pluralistic practices, in particular the use of constructive discursive methods rather than competitive voting for decision-making. Quakers have supported new egalitarian and pluralist movements, notably through *A Quaker Action Group*, set up in Philadelphia in 1966 to organize non-violent direct action on the Vietnam War. In 1971 the organization developed into the *Movement for a New Society* which was active until 1988 and developed tools and guidance for activism for nonviolent social change (Coover et al 1977).

Liberal Quakers are unusual among religious groups in that we have no orthodoxy—i.e. there is no expectation that Friends should have a shared belief system. However, Quakers do have an orthopraxy: a shared set of spiritual, organizational and decision-making practices and moral values by which Friends seek to live. This orthopraxy is a tradition in that values and practices are inherited and carried on, but it is continually and explicitly reconsidered in 'yearly meetings' which all members can attend. Yearly meetings form the top level of organization and decision-making for Friends. Quakers in most countries have one yearly meeting, but in others (in particular the United States) there are several, which may differ considerably in their traditions.

Britain Yearly Meeting has Christian roots, uses Christian language, and is sometimes called a 'church', but many British Friends do not consider themselves Christian. Quaker religious practice centers on 'meetings for worship' in which we sit together, usually for an hour, mostly in silence, during which participants may speak. Although the traditional idea is that these contributions, known as 'ministry', should be 'moved by the Spirit', Quakers do not necessarily ascribe to a belief in God or worship anyone or anything.

Liberal Quakers generally take both scripture (the Christian Bible) and past Quaker writings as sources of inspiration rather than absolute authority. Quaker practices are underpinned by a number of essential experiences and insights but at the centre lies a listening spirituality. The listening is both inward:

> Take heed…to the promptings of love and truth in your hearts

and outward:

> Receive the vocal ministry of others in a tender and creative spirit. Reach for the meaning deep within it… (Britain Yearly Meeting 1995)

Many people say they attend Quaker meetings because they 'feel at home in the silence'. There is no particular instruction or expectation of what people will do in their heads in the silence, although advice is available and experiences of meeting for worship are shared and discussed in a variety of ways. Many find a home in Quaker meetings who felt they were rejected or judged in their previous faith groups because they were different in some way. The Society of Friends is a place where individuals can be themselves and express their own truth, provided they adhere to the central Quaker practices.

On the other hand, Quakers speak of a 'gathered meeting' as one in which a kind of collective consciousness has developed. There is a feeling of the meeting as a presence, and that the spoken contributions are made on behalf of the whole meeting rather than the individual speakers. It is as if the meeting itself is becoming self-aware—or at least participants are becoming aware of the collective consciousness of the meeting as something that they are part of, and contribute to shaping.

One of the most important Quaker practices is our decision-making or 'discernment' method, called the Quaker business method. The traditional understanding is that the business method is the way Quakers find 'God's will' for them. However, many Friends are more comfortable with the idea that the purpose is to discern the 'leadings of the Spirit' or the 'sense of the meeting'.

Business meetings can be just about any size—from three or four Friends in a local meeting to 1200 or more in the sessions of Britain Yearly Meeting, the annual gathering of British Friends. In essence they are meetings for worship with an agenda. Small business meetings are sometimes quite informal, while larger gatherings need a formal discipline to work well. A 'clerk' invites spoken contributions on each agenda item. In more formal meetings, participants rise to indicate that they have a contribution to make. The clerk calls one of those standing to speak. Following each contribution there is a short period of silence before the clerk invites contributions again. Once the 'sense of the meeting' seems to be emerging, the clerk composes a draft minute seeking to capture it, and offers the draft to the meeting. Participants can then rise to make comments on the draft. If an individual has a serious doubt about the minute, there is usually an opportunity to express it, but Friends are expected be careful of attachment to personal positions and accept the sense of the meeting. Meetings sometimes have to record that they have been unable to reach unity, in which case a decision cannot be made. However, the process is certainly not perfect and decisions sometimes result from the preconceptions of the clerk, deference to the loudest or most respected voices, or a disgruntled minority giving way to the majority.

Some aspects of the Quaker business method—e.g. not voting—are shared by the consensus methods increasingly used in activist movements and other bodies (Seeds for Change 2013; WCC 2005). However, the Quaker business method differs from what is commonly understood as consensus in that it is not strictly egalitarian; it is not about participants agreeing intellectually with each other; nor is it a process of argument or negotiation between positions. It involves a spiritual process of self-forgetting, letting go of personal positions, listening deeply to each other and the Spirit (although participants may have different understandings of this), and sincerely seeking unity in recognizing the right way forward.

Quaker spiritual and decision-making practice is relevant to the sustainability challenge for several reasons:

- Quaker practice offers a particular blend of hierarchy, individualism and egalitarianism that seems to work in building community around difficult and complex decisions. It includes and moves beyond individualism rather than seeking to suppress it. As such it is particularly relevant to the need for a new kind of community in our individualized society.
- It nurtures a culture and discipline of constructive discourse and listening, enabling participants to achieve a shared understanding of a complex situation and of other people's positions and perspectives.
- It can develop solidarity around a collective way forward. It establishes (and relies on) a balance between the creativity and 'leadings' of individuals and the normative, 'testing' role of the community.
- Friends have developed, practiced and adapted our decision-making methods over several hundred years. Through this long experience, we have a thoroughly articulated understanding of the structures and personal disciplines involved in successful participatory processes.
- Quaker processes are readily transferrable to secular groups because they do not depend on participants adhering to any particular belief system or ideology.

Quakers use the business method to reflect explicitly on, and to develop, our values and ethics. The best established of these are often described as the Quaker 'testimonies' on peace, truth, equality and simplicity. Commitment to the testimonies is usually part of the conversation with new applicants for membership. Of these four, the peace testimony is the best known. It has led Friends to campaign against wars, refuse conscription, withhold taxes and take direct action to obstruct military activity. Quakers seek to express the testimonies on truth and equality primarily in our personal lives and in the functioning of our institutions, although we have also undertaken campaigns; e.g., for truth and integrity in public affairs. Simplicity is seen almost entirely as a principle for Friends' personal behavior, although most feel unable to live up to it.

3 Quakers and Sustainable Living

Quakers in Britain began to express concern about the environment in the 1960s, mostly seeing it as an extension of established values of nonviolence, equality and simplicity. Quakers have been strongly represented in the green movement in

Britain and have especially influential in the Green Party. Quaker Green Concern was set up in 1986 as an informal interest group for British Friends, with occasional gatherings and a regular newsletter. Various projects and committees on sustainability were set up by the Yearly Meeting, but they were mostly short-lived.

In the late 1990s Suzanne Finch, working as an individual Friend with a 'Rowntree Fellowship' from the Joseph Rowntree Charitable Trust (JRCT), visited meetings around Britain facilitating workshops on sustainability (Finch 2000). Her work culminated in a Yearly Meeting session in 2000 asking whether there should be a Quaker 'testimony to the Earth'. The session did not reach a strong conclusion and at the end of the Yearly Meeting participants in the children and young people's program expressed their frustration at the adults' indecisiveness. The matter was put back on the agenda of Yearly Meeting in 2001, which produced a 'call to action' to Quakers but no specific commitment.

In 2002 I set up the Living Witness Project, again working independently with a grant from JRCT, to develop a network of Quaker meetings responding to the Yearly Meeting call to action. From 2002 to 2008, Living Witness was the main support for Quaker meetings in Britain to explore and develop their approach to sustainability. It worked closely with Quaker Green Concern (renamed 'Quaker Green Action') and the two organizations merged in 2008. Living Witness has provided workshops for about 200 Quaker groups, supporting them in finding their way forward in acting on sustainable living. It maintains a network that currently includes about 100 local meetings and 500 or more individual Friends with twice-yearly national gatherings, summer schools and a variety of printed and Internet resources.

From 2008, other Quaker organizations in Britain and elsewhere initiated programs to develop their approaches to sustainability and to support Friends and meetings. Then, in 2011 at a week-long gathering in Canterbury, Britain Yearly Meeting considered sustainability in depth, with a number of presentations and lectures on the theme (notably Lunn 2011). The Yearly Meeting made a commitment on behalf of its 14,000 members to become a low carbon sustainable community (the 'Canterbury Commitment'). Local meetings were encouraged to work with a *Sustainability Toolkit* (LW/QPSW 2011) and to report on their greenhouse gas emissions and the actions they were taking to reduce them.

The Canterbury Commitment is one of the strongest and most far-reaching decisions that Britain Yearly Meeting has made for many years. To the extent that it is implemented, it affects not only the practices and priorities of local meetings in Britain, but also the lives of individual Friends. However, the process of implementing the commitment seems likely to be slow and painful. Despite the apparent unity on the general principle that climate change is happening and demands our attention, Friends have many diverse approaches. Among those most strongly committed to the concern there are some who advocate and are committed to a green, Earth-centered spirituality. Others have focused more on the practicalities of reducing their impacts on the environment, whether through frugal living or investing in new technology. A small number of very vocal Friends feel we should be paying attention to population rather than lifestyle.

Quakers are reluctant to suggest or acknowledge that we might have anything particular to bring to the sustainability agenda. Some have given up on the multiplicity of Quaker approaches and are focusing their energy on working with Transition initiatives, the Occupy and other activist movements, or other faith groups. The Canterbury Commitment was framed in terms of the moral imperative to reduce greenhouse gas emissions, rather than any particular Quaker contribution.

There is also a large proportion of Friends who see climate change as important but do not see reducing emissions as a priority or a possibility. This may be a form of denial (Weintrobe 2013) or there may be truth in the position. Some think it is too late, and there is growing interest in work on resilience, looking at how we can support our local communities to survive disaster. Some Friends with a strong personal spiritual discipline see climate change as a matter for acceptance, and for exercising non-attachment. Some Quakers feel that they have other, higher moral priorities—such as working to relieve specific instances of suffering and injustice—which often means they are committed to international air travel. And of course there are many who feel unable to change—saying they are too dependent on their car, cannot give up flying to see their loved ones, or cannot turn down the heating because they are sensitive to the cold.

Work is ongoing to clarify what the Canterbury Commitment means and to support local Quaker meetings in engaging with it, as well as to document and celebrate what is already being done. Some frustration is being expressed at the lack of focus and of a common understanding of the challenge. Yet, despite the diversity of personal positions, Quakers do seem to be taking action together. Individual Friends have found much more support for greening their lives than they did prior to the Canterbury Commitment. Local meetings are making considerable investments to green their buildings. Other Quaker organizations in Britain are proactively reducing their carbon footprints and otherwise seeking to be public examples of sustainability. For example, the restaurant at Friends House (the national headquarters) won a 'sustainable restaurant of the year' award.

4 Conclusions

British Quakers have come a long way towards developing a coherent approach to sustainability. There is no single shared narrative about what matters, or what should be done about it, and there probably will never be one. Yet many Friends are taking significant steps towards sustainable living and see it as part of their shared identity as Quakers. The Quaker process represents an interplay of individual initiative and group discourse, working with tradition and hierarchical structures, that may offer insights for others working in this field.

The Quaker approach is grounded in a spirituality that focuses not so much on the human relationship with nature, as on our relationship with each other. Open listening is a central discipline. The Canterbury Commitment includes encouragement to be tender with each other, and to ensure that our action arises from love. The

Quaker commitment to 'answering that of God in every one' means that apparently conflicting positions are accepted and included.

The diversity of personal positions among Quakers is small compared with that in wider society. Most Friends are conscious of being out of tune with consumer culture, yet we are all deeply infected with it and it is reflected in the difficulty most of us find in limiting our own consumption. Few Friends are ready to abandon the central heating or the car—although they will put solar panels on the roof and turn down the thermostat a little, and Quaker cars are usually on the small side.

Quakers are one of the groups in society who might be expected to be most willing and able to move towards sustainable living. The needed shift is happening, but it is contested and it seems to many of us to be too slow.

Perhaps the main hope for humanity is that many examples are being developed, by Quakers and others, both of practical initiatives for sustainable living and of the spiritual and social practices that underpin them. These examples may prove invaluable if the world does finally wake up to the reality and urgency of climate change.

References

Baudrillard, J. 1970. *The consumer society—Myths and structures (reprinted 1998)*. London: Sage.

Beck, U. 1988. *Risk society: Towards a new modernity*. London: Sage.

Berners-Lee, M., and D. Clark. 2013. *The burning question: We can't burn half the world's oil, coal and gas, so how do we quit?* London: Profile Books.

Britain Yearly Meeting of the Religious Society of Friends 1995. *Quaker faith and practice*. London: BYM.

Burgess, J. 2003. Sustainable consumption: Is it really achievable? *Consumer Policy Review* 11 (3): 78–84.

CE [Council of the European Union] 2009. Council Conclusions on EU position for the Copenhagen Climate Conference (7–18 December 2009). http://www.consilium.europa.eu. Accessed 30 July 2013.

Coover, V., E. Deacon, C. Esser, and C. Moore. 1977. *Resource manual for a living revolution*. Philadelphia: New Society Publishers.

Douglas, M., and S. Ney. 1998. *Missing persons: A critique of personhood in the social sciences*. Berkeley: University of California Press.

Finch, S. 2000. *A Quaker testimony to the Earth?* London: Friends House.

Giddens, A. 1994. Reflexive institutions. In *Reflexive modernization*, ed. U. Beck, A. Giddens, and S. Lash, London: Polity Press.

Hopkins, R. 2013. *The power of just doing stuff: How local action can change the world*. Cambridge: Green Books.

Howell, R.A. 2012. *Promoting lower carbon lifestyles*. Doctoral dissertation, University of Edinburgh, UK.

Jaeger, C.C., O. Renn, E. Rosa, and T. Weber. 1998. Decision analysis and rational action. In *Human choice and climate change. Volume 3. Tools for policy analysis*, eds. S. Rayner and E.L. Malone, 141–215. Columbus: Battelle Press.

Kahane, A. 2004. *Solving tough problems: An open way of talking, listening, and creating new realities*. San Francisco: Berrett-Koehler Publishers, Inc.

Loring, P. 1999. *Listening spirituality. Vol. II: Corporate spiritual practice among friends*. Washington,DC: Openings Press.

Lunn, P. 2011. *Costing not less than everything: Sustainability and spirituality in challenging times*. (Swarthmore Lecture, 2011). London: Quaker Books.

LW/QPSW [Living Witness and Quaker Peace and Social Witness] 2011. *Sustainability toolkit: Becoming a low-carbon, sustainable community*. London: Quaker Books.

Michaelis, L. 1998. Economic and technological development in climate scenarios. *Mitigation and Adaptation Strategies for Global Change* 3 (2–4): 231–261.

Michaelis, L. 2008. Quaker simplicity. In *Frugality: Rebalancing material and spiritual values in economic life,* ed. L. Bouckaert, H. Opdebeeck, and L. Zsolnai. Oxford: Peter Lang Academic Publishers.

Rostow, W.W. 1990. *The stages of economic growth*. Cambridge: Harvard University Press.

Schumpeter, J. 1943. *Capitalism, socialism and democracy*. Cambridge: Harvard University Press.

Seeds for Change 2013. *A Consensus handbook: Co-operative decision making for activists, co-operatives and communities*, Seeds for Change Lancaster Co-operative Ltd. http://www.seeds-forchange.org.uk. Accessed 12 July 2013.

Thompson, M., and S. Rayner. 1998. Cultural discourses. In *Human choice and climate change. Volume 1. The societal framework,* ed. S. Rayne and E.L. Malone, 265–343. Columbus: Battelle Press.

WCC [World Council of Churches] 2005. *Manual for conduct of meetings of the world council of churches*. http://archived.oikoumene.org. Accessed 30 July 2013.

Weintrobe, S., ed. 2013. *Engaging with climate change: Psychoanalytic and interdisciplinary perspectives* London: Routledge.

Laurie Michaelis co-ordinates Living Witness, a charity working with Quaker and other groups in Britain to develop understanding of and action for spirit-led sustainable living. From 1998 to 2002 Laurie worked at the Oxford Centre for the Environment, Ethics and Society as research director for the Oxford Commission on Sustainable Consumption. Other projects have included a study for the European Environment Agency on consumption futures in Europe and an assessment of the policy implications of sustainable consumption research for the UK Sustainable Development Commission. He previously worked as a policy analyst at the Organization for Economic Co-operation and Development (OECD); and as an energy technology analyst at the UK Energy Technology Support Unit. He has been a lead author and convening lead author for several reports of the Intergovernmental Panel on Climate Change, including the Special Report on Emissions Scenarios.

Sustainability and Long-Term Growth in the Financial Market System

Aloy Soppe

The paper discusses the implications of the unequal rate of growth of the international financial market on the one hand and the growth of the real economy on the other. Based on viewing balance sheet equilibrium as a prerequisite for sustainable economic growth, a discussion is presented about the optimal interest rate, which depends on the natural and real economic growth of the economy. In the final section, two propositions are presented. The first proposition is that the average interest rate should vary with the rate of growth of the real economy which is a proxy for organic growth in economic processes. Lower interest rates are not desirable because of their impact on economic opportunities. Higher interest rates should be rejected because of the implicit additional positive time preference that they entail. The second proposition concerns the accumulation of wealth. Under the assumption of the existence of a positive interest rate, the absence of a financial tax and the absence of bankruptcies, financial capital grows exponentially and therefore needs to be managed institutionally, in a positive way. A structural imbalance between the monetary sector on the one hand, and the real economy on the other, leads to the illusion of purchasing power in the hands of the public.

1 Introduction

The cessation of the convertibility of the US dollar to gold in 1971 by President Nixon can be considered a landmark event in the modern global economy. By depegging the dollar from gold, all the world currencies officially started floating, and from then on structural changes took place in the financial industry. Now, 40 years later, we can see that the crash of 1987, the internet bubble of 2001, the many revelations of financial fraud at the corporate and individual level and the banking crisis

A. Soppe (✉)
Erasmus University Rotterdam, Rotterdam, The Netherlands
e-mail: soppe@law.eur.nl

L. Zsolnai (ed.), *The Spiritual Dimension of Business Ethics and Sustainability Management*, DOI 10.1007/978-3-319-11677-8_10

of November 2008 are clear indications of the vulnerability of the global financial system that was developed and built in the last decades of the former century.

Three major developments are regarded as the drivers of modern financial culture in the Western world. Firstly, the *communications revolution*, as initiated by the Internet, that globally facilitated the financial sector to change its culture from service-driven to a leading driver of economic growth. The banking and investment industry, for example, used to be a 'gentlemen's business' that primarily served its clients with high standards of honesty, integrity, care and diligence. However, in the profession over the last two decades entry barriers have disappeared and key employees are now typically creative, risk-taking, success-driven, analytical and intelligent people who lack training in the common ethical values of the profession (Caccese 1997). Making money for the sake of making money has become a well-accepted goal.

A second crucial aspect in the changing financial environment is the *securitization process* of international financial markets. During the last decades, the importance of transparent traditional bank lending instruments has gradually declined in favor of anonymous market-traded assets. Important in that financial cultural change was the influence of the (increasing) role of institutional investors, hedge fund activities and the necessary risk diversification and reinsurance processes. Closely related to this trend was the substantial rise in the trade in off-balance-sheet products, derivatives and the development of program trading. The inherent volatility of underlying assets prices required an adequate set of financial instruments with which to manage financial exposure. Goldman Sachs became an iconic example of product development, financial culture and market manipulation.[1]

A third important factor is the growing influence of *civil society* regarding responsibility for social welfare. In the last decades, Europe in particular has been characterized by its implementation of sizable industry-driven privatization projects in fields of operation that were traditionally in the hands of the public sector. The increase in private competition has created a competitive advantage for the private sector over the public sector—with the latter gradually becoming a judicial infrastructure and more or less a 'lender of last social resort'. At the same time, we have seen that sustainable development has become increasingly important to modern companies as a response to customer trends and the need for reputation-related discipline.

The objective of this paper is to provide a long-term analysis of financial markets in the developed economy. After the collapse of communism—in which corporate control was dominated by the production factor of labor—the question at stake right now is whether financial capitalism will be able to survive. Today, it is the production factor 'capital' that dominates corporate control. Do we not perhaps need alternative and more democratic corporate models in order to take care of things like the interests of the natural environment (the third major production factor), and other stakeholders?

[1] See, for example, the CDO deal 'Abacus 2007-AC1', the subject of court proceedings initiated by the SEC in April 2010.

The approach that has been chosen for this paper is based on a moral-philosophical foundation that covers a wider analytical terrain than modern economic theory, which is based on a strict utilitarian foundation. The inherently normative approach integrates principle and consequence-based ethics. The primary function of money and financial capital—its role as a unit of account and exchange—is in danger of becoming subordinate to the creation of financial wealth as a goal in itself. Collective trust in financial markets, the major issue in finance, is at stake. The very moment that the neutrality of money and long-term capital is attacked by moral hazard and adverse selection, the financial system loses its scientific objectivity. The structure of the paper is thus as follows: Section 2 provides a short introduction of an alternative philosophical approach into the subject of interest and money. Given that, in practice, interest payments have facilitated real economic processes for centuries now, it is essential that modern developments are also examined against this background. The following sections translate the previous theoretical discussion into two basic propositions that relate to the current financial system. These propositions are subsequently explained and justified. Finally, a summary and conclusions are presented.

2 Philosophical Background

Our financial and economic analysis begins by asking where value comes from.[2] Since economic life was strongly allied with religious life in the Middle Ages, this question really only became relevant from the time of the introduction of division of labor. A natural product can only take on economic value if it undergoes some process that is carried out by people (e.g. coal and oil only have a value once they have been mined). Capital goods subsequently come into being as a result of human ideas and action. In other words, capital goods are chiefly created through intellectual work. These capital goods become productive because a decreasing amount of labor produces an increasing quantity of goods. The actual abstraction process that takes place here results in capital goods becoming productive without requiring any labor. And, because capital goods can be traded, their value can ultimately be expressed as a sum of money: assets. In this sense, assets are the realization of intellectual thinking. Even though capital goods can be considered to be the realization of past intellectual thinking, their value is determined through their application by the present owner of the capital goods. In other words, the value of assets is determined by a combination of intellectual activity that is realized, and future events.

Rudolf Steiner (1922) argued that the division of labor has caused the physical exchange economy to evolve into a money and skills-based economy. The economic system has explicitly absorbed the legal system due to the very existence of money

[2] This analysis draws heavily on the work of Rudolf Steiner (1861–1925). He is given a prominent place in this paper because his original economic thinking presented in the publication *National Ökonomischer Kurs* (Steiner 1922) forms the foundation for my stance concerning the sustainability of money and capital in the following sections.

itself. It is a fact that money can be considered to be a right to future purchasing power and, as such, takes on an importance equal to that of the production factors of nature, labor, assets and capital goods. Steiner distinguishes between three types of money: buying money, lending money and gift money. Buying money is the counterpart to commodities and is only used to trade and to gain diverse commodities. Lending money is created with every new direct investment. This is young and productive money which entails economic improvements, allowing for the generation of short term revenues as induced by human ideas and actions. Then there is gift money. This money is needed for people who are not able to produce commodities, like children, pensioners or sick people who cannot work. This money is needed to create social value which is not a commodity, such as education, or the work of artists, scientists, religion, etc. A basic problem in the economic analysis of the financial sector is its inherent lack of sustainability. Steiner reasons from a closed world-economic model and from a general universal law that is considered applicable to economic life as a whole.

Based on this 'universal law', Steiner goes on to suppose that all material objects such as consumer goods, capital goods and services age continuously and therefore intrinsically diminish in quality and value—in contrast to money! Money represents a right to future purchasing power with an essentially everlasting quality, which implies that in the course of time money acquires too much value in relation to goods and services. At the same time, interest payments lead to social injustice (in terms of distribution of income), because the quality of material goods and services can only be maintained through labor while the interest mechanism causes capital to increase in value (even exponentially) without the need for labor. This inevitably results in the accumulation of value, something that also occurs, according to Steiner, with the value of land and capital goods.

So how does Steiner suggest that this unbalanced situation of growth be solved? Loan capital offers the opportunity to achieve economic growth. Lending money gives the right, as it were, to a service in return. So Steiner goes on to define interest as that which one receives if one renounces a service in return. In this sense, the interest represents the dissolution of the physical exchange that is veiled behind a monetary transaction (caused by the division of labor). In other words, the forces of the conflict are transformed and bundled in the interest. To compensate for this, contra-forces must be brought into play. In Steiner's theory, these united forces are represented by associations. Products, consumers and traders are parts of these associations. Associations are parties who together play a subtle power game and who ensure that the primary survival instinct and greed of organisms are finely tuned. According to Steiner, the accumulation of value in land and capital goods mentioned above must now be creamed off by means of gifts from economic actors to the field of culture (which in fact results in destruction of capital). The association is tasked with managing this process. If capital is not destroyed voluntarily through gifts from traders, then depreciation will occur involuntarily in the form of inflation and bankruptcy. There should therefore be a continuous and institutional transfer of value from the economic order to the legal system and the cultural order. Through the gifts, capital will acquire a temporary character (just as in real business cycles) and will cease to exist if the recipient has consumed it.

In my opinion, Steiner, with his encouragement of gifts, combines a Greek philosophers' approach to the problems of redistributing wealth with Jewish and Islamic ethics about interest-free loans. Interest-free loans made in a close circle can also be considered gifts. However, there is one great difference. Steiner makes a distinction between money for purchasing, money for lending and money for giving. Money for purchasing belongs to the economic order and may not be given interest free because this would redeem or reimburse, as it were, the underlying realistic exchange transaction. But money for giving has a twofold nature. On the one hand, such a transaction is initiated in economic life in order to facilitate cultural life. On the other hand, it prevents value accumulating in land and in excess capital.

3 Propositions for Sustainable Finance

An important question is whether the culturally, historically and often religiously-based discussion regarding the appropriate level of interest—including the possible risk of capital accumulation—is still relevant to modern times. The financial sector dominated the global economy during the period from 1970–2012, with, admittedly, a positive impact on income and employment, but with a clearly negative impact on global and local financial stability. The market interest rate as an information carrier has become indispensable in modern economic processes. Interest payments and interest rates in current financial markets signal both the financial power and weaknesses of nations (with the yield curve as the intertemporal price differentiation). The essential task of interest rates is to enable a flexible allocation of financial assets between buyers and sellers on the open market. The existence of successful real economies without well-functioning financial markets is hard to imagine. The historical and ethical approach to interest payments—as presented in the former sections—forms the ethical-introductory framework for the coming analysis on the long-term equilibrium between real and financial markets in the modern market economy.

In this paper I argue that *capital global growth rates should be similar to rates of real economic growth processes*. As an extension of the social discount rate (Turner et al. 1994), extreme interest rates (either too low or too high) are interpreted as ethical premiums because of the assumed sustainability premise that long-term financial claims should equal long-term physical assets. From that perspective, the interest rate does not just represent the market price of capital which depends on demand and supply, but is a variable with important implications for the welfare economy.

The Role and the Optimal Level of Interest Rates

The basic theory that justifies capital markets in the real economy is expressed by the separation axioms of Fisher (1930). If we assume that markets exist under conditions of perfect certainty, financing decisions can be taken independently of investment decisions. The introduction of capital markets increases the utility of both the agents with an abundance of wealth (savers) and the usefulness of agents with investment opportunities (entrepreneurs). The integration of real trade theory and the theory of finance markets under certainty and uncertainty is described by MacDougall (1960). Later, these models were refined and expanded to include risk-free bonds (Grossman and Razin 1984) and international markets (Hodder and Senbet 1990). As a result, the necessity of capital markets is widely accepted from a utility perspective, which leads to the next question: if we need capital markets in the market economy, what then is the optimal interest rate from a social and ethical perspective?

In this paper, two major propositions are advanced. The first proposition is rooted in historical explanations for the charging of interest based on the two main motives in the literature; namely, the opportunity cost of capital and time preferences–as described by Fisher (1930).

Proposition 1 *In a sustainable financial system, the average interest rate needs to be equal to the rate of real economic growth in the long term.*

This proposition is derived from the well-known quantity theory of money by the Monetarist School. It states that the quantity of money should equal the quantity of real economic production:

$$\mathbf{M} \times \mathbf{V} = \mathbf{P} \times \mathbf{T} \tag{1}$$

where, during a fixed period:

M = Total amount of money in circulation on average in an economy
V = Velocity of money circulation (the number of times money changes hands)
P = Average price level of goods and services
T = Volume of transactions of goods and services in an economy

The point of departure is the monetarist requirement that there is short-term balance sheet equality between the monetary sphere on the one side and the real economic sphere on the other. The purpose of the monetary market balance sheet equation provided by the monetarists was to show that, in order to control inflation (represented by P), the growth rate of the supply of money (M) needed to be balanced and, if necessary, restricted, depending on the money velocity (V) and growth in real trade (T). During the 1980s and 1990s, the FED published a weekly overview of the targeted money supply in the United States of America in an attempt to manage inflation.

Building on a similar concept of balance sheet equality, it can be stated that, as with the supply of money, growth in capital (financial claims with a maturity of longer than 1 year) also needs to be in line with the long term growth rate of the physical economy. The short term equality of the monetarists is thus converted into a long-term balance sheet equality of sustainable economics (Fig. 1).

Real Assets		Monetary Assets	
Physical Infrastructure		Stocks	
Buildings		Bonds	
Machinery, etc.			
Human capital (number of educated people)			
∑ Physical Assets	PA	∑ Monetary Assets (claims)	MA

Fig. 1 Balance sheet of a sustainable economy

Splitting the sum of the physical assets and the monetary claims into a volume and an average price level component, we get:

$$\mathbf{V_{PA} \times P_{PA} = V_{MA} \times P_{MA}} \tag{2}$$

where:

$\mathbf{V_{PA}}$ = Volume of all physical (real) assets
$\mathbf{P_{PA}}$ = Average price level of physical assets
$\mathbf{V_{MA}}$ = Amount (volume) of capital; capital is defined as financial claims with a maturity of 1 year or longer
$\mathbf{P_{MA}}$ = Average return on stocks and bonds

Expressed in words: the value of real economic assets should equal the value of financial long-term claims (capital). This balance sheet equation is based on the premise that if the value of financial claims increases faster than that of real assets, trust in the financial sector diminishes because the proportion of real economic assets to financial claims becomes too small. Long term income can only be generated from real economic projects and not from trading financial claims. From this balance sheet equation it can subsequently be derived that, to maintain a stable ratio between real and financial assets, the levels can be converted into growth rates. We thereby obtain:

$$\mathbf{g_{pa} + p_{pa} = g_{ma} + p_{ma}} \tag{3}$$

where

$\mathbf{g_{pa}}$ = Volume growth of physical assets
$\mathbf{p_{pa}}$ = Price change of physical assets (inflation)
$\mathbf{g_{ma}}$ = Volume growth of monetary assets (accumulation)
$\mathbf{p_{ma}}$ = Price change monetary assets (interest rate)

Expressed in words: by making the static balance sheet equality dynamic, the growth rate of the longer-term monetary economy (the average interest rate: p_{ma}

plus the volume change of capital claims: g_{ma}), should equal the growth rate of the real economy in a sustainable economy (the change of the average price level of assets p_{pa} plus the change in volume of physical assets g_{pa}). If we assume that the change in volume of capital and the change in volume of real assets remains constant (growth is zero), proposition I states that the change in the interest rate (p_{ma}) should equal the change in the price of real assets (p_{pa}).

The questions at hand right now are 'what is the optimum interest rate?', and 'what is the relevant rationale which may be employed to explain sustainable interest rates, not only from an economic perspective?' Many dimensions and factors are involved in the setting of interest rates; a few are mentioned below. However, a positive rate is desirable; a proposition which is supported by the following arguments:

i. *The opportunity cost argument*. The opportunity cost in this context means the productivity of the inherent production capacity in the real economy. Starting with the classic labor theory of value—which states that only labor is value enhancing in the production process—a positive interest rate is needed to explain the opportunity cost. Then, by adding the value-creating energy of entrepreneurship and acknowledging the creative role of humans in our economy we may state that a positive interest rate is needed in the sustainable global economy. Thoughts and ideas from the past accumulate in capital goods that facilitate more productive future production and are represented by an independent value given in the form of capital. Capital goods are thus realized human creativity from the past. In other words, physical labor, entrepreneurial ideas and technological innovations of the past accumulate in a value called 'capital goods' at $t=0$, which continues to produce value into the future. Note that the values according to this view are therefore not determined by future cash flows that are capitalized, but by the inventiveness and creativity (perhaps even spirituality) of the realized concept of the past. The natural vigor of the economic process justifies a positive real interest rate in order to financially facilitate economic progress. This ex-post driven growth component is thus at odds with the conventional wisdom that economic value is only determined by discounted future cash flows.
ii. *Fee earning capacity*. In order to finance the entire banking sector, a positive interest margin is required (the transaction costs argument). A positive interest rate, from this perspective, is helpful. However, a positive interest rate is not a prerequisite. Even with negative deposit rates, it is possible to create positive interest margins.

In addition to these arguments for positive pressure on interest rates, the traditional upward pressure on interest rates that results from a pure liquidity preference is less desirable from an ethical point of view, because:

a. Based on the principle of *equality between generations,* an interest rate of zero can be argued for (assuming no time value difference between time periods). There are as many reasons for a negative as for a positive time preference. The choice in this respect reflects the normative moral stance and may be far from a positive value (see the discussion in Sect. 3.1 and the six traits of time preference as formulated by Fisher (1930, 1907)). Pure time preference is, I believe, best described by Fisher's 'impatience to consume'.

b. Starting with the premise of a sustainable society, intertemporal analysis of risk is less relevant since time as a variable is considered infinite and consequently cut off from analysis. This has major implications for the modern concept of risk. Risk is an accepted phenomenon in investment theories as the explanation for premiums in the yield (interest), but if time is lost as a source of uncertainty, the risk decreases. Lower- than-anticipated growth is something against which one cannot hedge, but something that one should always be prepared for—the same is true in a moral sense. Risk management and other ways of fixing future cash flows are therefore less important.
c. High or too high interest rates—as often found in the current market model—frequently lead to negative consumption effects:
    - Keynesian analysis indicates that a high interest rate negatively affects the real economy because it promotes more capital expenditure on real investment projects.
    - Based on the concept of duration it has been found that high interest rates lead to lower interest rate sensitivity. In other words, the higher the interest rate, the lower the indifference of investors with respect to interest rate variability.
    - Inflation and overheating of real economic growth is, in modern monetary policy, contested by an attempt to tighten the money supply by raising interest rates. This tightens future growth because investment decisions are then only correct if the anticipated growth at $t = 0$ is actually realized. If at $t = 1$ the realized growth is lower (because the market rate, for whatever reason, has fallen), the net present value analysis has failed, just as has the inter-generational decision criterion at $t = 0$. The types of ex-ante analyses that prevail in current financial calculations are, in my opinion, only a partial theoretical justification for an optimal rate.
d. Higher interest rates lead to greater income inequality. Empirical research shows that of all the countries in the world in the period from 1948 to 1993, roughly 80 % paid interest to the remaining 20 % (Van Schaik 1996, p. 61). Krueger and Perri (2006) find that consumption inequality is much less than income inequality. The difference can be explained by the fact that the poor and middle classes accumulate more debt to support their consumption, thus keeping consumption inequality in check, while the rich invest their savings in assets backed by loans to the poor and middle classes. This eventually leads to even higher income inequality (Kumhof and Rancière 2010).

In summary, it is argued that the real interest rate functions optimally when it varies in tandem with the real growth rate of the economy, which is a natural proxy of organic growth in the economic process. Lower interest rates are not desirable from the perspective of the economic idea of opportunity. Higher interest rates should be rejected because of the implicit additional positive time preference that they entail.

Capital Accumulation

A second proposition focuses on the volumes of physical assets and the volume of capital claims with a maturity of longer than 1 year. This is expressed using the symbols $V_{PA} = V_{MA}$, assuming that ΔP_{PA} and ΔP_{MA} are zero. Just as physical capital goods and immovable have a natural tendency to wear, capital claims also need to have a finite maturity. If not, the volume of capital claims grows exponentially due to the compound interest mechanism.

Proposition 2 *In a sustainable financial system, capital accumulation needs to be prevented through institutional capital destruction and charity in order that its volume of growth is the same as the volume in growth of physical goods.*

Although interest rates are widely considered to be the price of money and capital, interest can also be interpreted as the growth rate of output. This forgotten dimension is only stressed by certain authors (e.g. Boulding 1992, pp. 79–80), but is essential for understanding the stability of the economic system. Real economic processes are driven by a natural process of wear and entropy. This occurs because of consumption and the physical processes of the wearing of all material things. To a consumer, this process is clearly visible (after a piece of bread is consumed it has completely 'gone'), but for durable capital goods such as buildings and other fixed assets, a longer period of time is involved. Nevertheless, the ravages of time make an impact and do their natural job. In the accounting profession, new financial assets are deemed to depreciate in order to account for this decline in the material integrity of equipment.

The monetary process is different. The essential difference between real economic processes and monetary processes is rooted in the character of the transactions involved in our monetary and economic exchanges. Where in real economic transactions, physical goods are exchanged for money or a claim on money, in financial transactions money is exchanged for a claim to money (see CBS 1983, p. 16). The difference now is that, because of the interest mechanism, a monetary contract is a claim on a monetary asset that intrinsically increases in value due to the expected return on that underlying, and possibly derivative, asset. This contrasts with the value of the physical goods in the real economic transaction. A commodity is consumed immediately after the transaction. For example, the legal financial claim on a newly-issued bond yields an ongoing potential cash flow after interest payments and repayment, assuming that the very same bond is permanently reinvested. If these rights, once created, are transferable to debt security, they may, in theory, retain their purchasing power quality indefinitely. For equity, this is not the case. Shares, as a classic example of capital, have a permanent maturity and the value of the share depends on the quality of the underlying, real, assets.

Many other examples of financial assets that are traded, such as CDs (Certificates of Deposit), CP (Commercial Paper), CDOs and CDSs etc., have, theoretically, an eternal quality as long as a positive net return is achieved and the initial loan is repaid. The primary nature of financial assets concerns future consumption, whereas in a real economic transactions consumption takes place immediately. The idea, described above, of the existence of exponentially-growing financial assets is based on three major quality conditions. Firstly, the absence of bankruptcy risk is assumed. When companies go bankrupt and debt will not be repaid, this has the same effect as an enforced gift from creditors to the company. In this case, capital is destroyed and the macroeconomic volume of capital decreases (in other words, capital is consumed). Secondly, it is assumed that no capital tax is levied. Capital tax is money earned by the government at the expense of the investor that will be spent (consumed) immediately in the economy. This also reduces the volume of capital because it cannot be reinvested immediately. Thirdly, it is assumed that a positive real interest rate is paid. If inflation is higher than the nominal interest rate, the purchasing power of capital decreases, implying once again destruction of capital.

Based on the previously-mentioned differences in character between real economic transactions on the one hand and financial transactions on the other, it is argued that the imbalance between the real and the monetary sector needs to be monitored and managed positively. Where capital goods have a natural tendency to wear and decrease in value, financial capital grows, in theory, exponentially through the compound interest mechanism. In practice, however, the value of capital is permanently endangered by inflation, moral hazard, defaults and taxes. By borrowing money from a bank (=one party's debt and another's capital), temporary purchasing power is obtained until the moment that the debt, including interest payments, is repaid. Repayment destroys the capital initially created by the loan, but the pre-existing capital may increase with interest paid. This process of borrowing and lending creates a permanent stream of income from debtors to lending parties, potentially leading to serious income differences (Fig. 2 is an illustration of the current European situation).

Why has the financial system not already gone bankrupt under the pressure of the rate of accumulation? The value of the accumulated capital is systematically destroyed and damaged by inflation, stock market crashes, bankruptcies and natural or other disasters. Wars are also an effective means of destroying capital. The paradox now is that economists do everything they can to prevent bankruptcies and inflation, but these events turn out to be necessary for the sound reduction of capital accumulation. A conservative attitude is socially desirable because disasters, bankruptcies, wars and inflation are economically, politically and socially regarded as negative. However, from the technical perspective, reductions in capital turn out to be valuable and necessary in order that there is balance between the monetary economy and the real economic supply of capital goods.

How then, can we tackle the accumulation of capital in an institutionally positive way? The normal way to destroy capital is by repaying debt. As we all know, this becomes increasingly difficult as the amount of debt increases (will Greece succeed?) However, another question arises: who is morally responsible for an institution that defaults? The borrowing institution, in an attempt to recover control, or the lending institution that receives higher interest payments because of the higher risks it takes? Apart from repayment of debt, there is only one unequivocally positive way to destroy capital and that is charity/donation. There are basically two forms of charity: private donations and redistribution via the government (taxes); the latter having a more or less forced character.

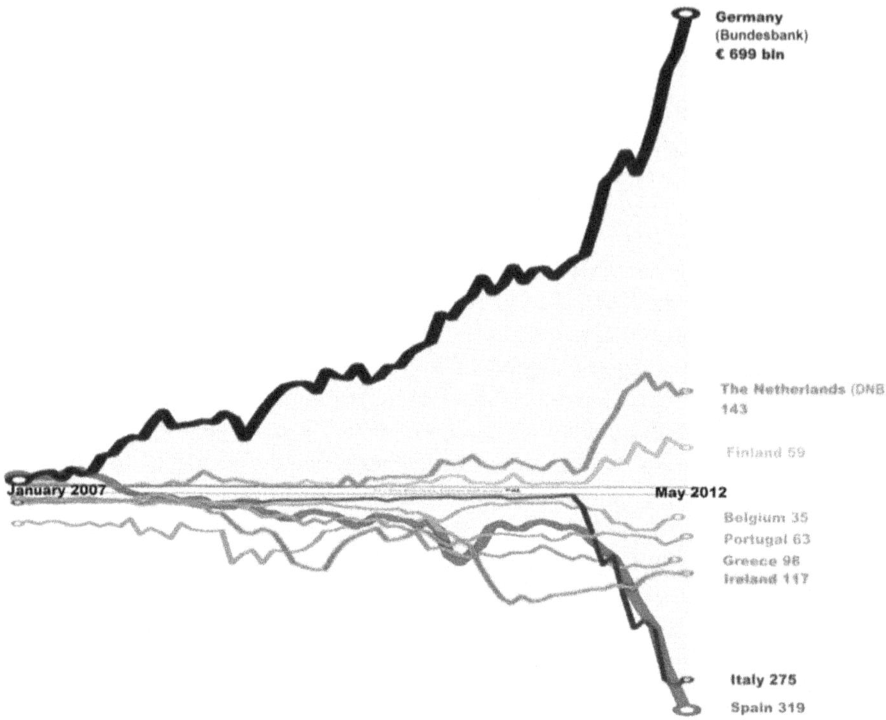

Fig. 2 Capital and debt of European Central Banks at the ECB. (Source: The Dutch daily newspaper 'Financieel Dagblad', 5 July 2012)

Let us start with donations. A donation is a gift that does not have to be repaid—capital is destroyed and the money is consumed completely. This is (economically) exactly what we need. Moreover, it leads to socially desirable effects. The donor feels appreciated and is economically free to nominate the recipient of their donation. The recipient is obliged to spend the money, which leads in a natural way to consumption or production in the poorest (or most vulnerable) sectors of the economy, such as music, the arts and non-commercial education and health care. The latter sectors, as we can all observe, are under the most pressure in the current economic climate.

Private donations, as described above, are the purest form of charity. Taxes that are designed to redistribute wealth between citizens can be seen as a more collective form of charity. Our current tax system primarily operates by taxing income and taxing spending (VAT). In this context, we should consider the utility of the proposed Financial Transaction Tax (FTT) in Europe. The potential tax revenue is enormous; trillions of Euros of revenue per year, worldwide. Assuming that banks act in a morally responsible way and do not increase the fees they charge clients, financial taxation enables governments to decrease the rate of growth of capital and increase spending on education and social projects.

Why is the destruction of capital now so crucial in order for financial markets and institutions to become sustainable? The key is that all participants are mortal in the real economy. People (and human capital) die and are reborn, and, like physical inputs (buildings, machinery, infrastructure), are permanently depreciated, demolished and rebuilt. Furthermore, if the inputs are of good quality, these impulses are funded with savings from the past; namely, depreciation. Why then does a debt title theoretically have an eternal life? Why not introduce gift bonds? For example, a 4 % regular 10-year bond may, in principle, be indefinitely reinvested and accumulate in volume exponentially. An alternative to this would be a 10 % ten-year bond which does not have to be repaid. In this case, only the principal would be repaid in the subsequent 10 years, without interest. This does not automatically increase the amount of capital; the initial loan can be consumed in 10 years. Is this irrational lending behavior? In itself, this behavior could be explained, for example, by a possible negative time preference for a lender with a shorter time horizon (Fisher 1930) or by Calvin's admonishments about restricting the raising of interest. Even better than the above example would be a 10-year 4 % bond, whereby only 400 € (for example) would be repaid after 10 years and 600 € of assets would be destroyed immediately at the time of issuance of the bond. Are these ideas unrealistic and utopian? Maybe. On the other hand, the time preferences of potential donor institutions are structurally different from those of today's dealers and speculators. The banking system after the crisis will hopefully not neglect the long-term balance equation between the real economic and the monetary sector, as described at the beginning of this section.

What is utopian, in my opinion, is unbridled faith in the current financial paradigm. The structural disparity between the real sector on the one hand and the financial balloon on the other hand is not sustainable. As in the past—the collapse of communism occurred primarily because of the unsustainability of the dominance of the production factor 'labor'—we are now waiting for the downfall of a system that floats on the dominance of the service factor 'capital'. A sustainable economy can only be created through the democratic and fair implementation of a market economy in which the production factors of labor, capital and the natural environment are of equal importance. The FTT tax has so far proved politically unfeasible. The morality of participants in the market economy and the absence of international regulation and supervision are to blame. Let us hope that the recent US and European financial crises lead to the insight that the financial sector needs be drastically reformed. A tax on financial transactions is a fair way to bring financial markets back into a balanced position.

In summary, we can say that the compound interest rate mechanism on the one hand and the lack of institutionalized destruction of capital claims on the other hand cause the economy to maintain an unsustainable rate of growth. In the goods and services economy a natural decrease in volume occurs (because consumer and capital goods age and are physically destroyed), which leads to a growing imbalance between the monetary sector on the one hand and the real economy on the other. This imbalance between the actual volume of production and the volume of capital remains statistically hidden as long as the world's total assets are insufficiently mapped and the issue is not policy-relevant to monetary authorities. Nevertheless, it should be clear that the structural differences between the monetary and the real economic sector lead to an illusion of real purchasing power. Proposition 2 is intended to signal this imbalance. Empirical verification is necessary but difficult due to the existence of large amounts of black and grey data on the international economy. This is nevertheless a topic for future research.

4 Conclusion

This paper presented a discussion about the implications of the decoupling of the rate of growth of the international financial market from the rate of growth of the real economy. Based on balance sheet equilibrium as a point of departure for sustainable economic growth, arguments were presented about the optimal interest rate, depending on the real economic growth of the economy.

The first proposition stated that the average interest rate should vary with the growth rate of the real economy, which is an estimator of organic growth in economic processes. Lower interest rates are not desirable because of the economic idea of opportunity. Higher interest rates are rejected because of the implicit additional positive time preference that this entails. The second proposition refers to the accumulation of wealth. Under the assumption of a positive interest rate, the absence of a financial tax and the absence of bankruptcies, financial capital grows exponentially and needs to be managed institutionally if a state of sustainability is to be achieved.

References

Boulding, K. E. 1992. From chemistry to economics and beyond. In *Eminent economists,* ed. M. Svenberg. Cambridge: Cambridge University Press.

Caccese, M.S. 1997. Ethics and the financial analyst. *Financial Analyst Journal,* 53(1):9–14.

CBS, 1983. Nationale rekeningen. *Centraal Bureau voor de Statistiek* (The Netherlands). Staatsdrukkerij.

Fisher, I. 1907. *The rate of interest.* New York: MacMillan Company.

Fisher, I. 1930. *The theory of interest—As determined by impatience to spend income and opportunity to invest it* (reprint). Clifton: M. Kelly Publishers (1974).

Grossman, G. M., and A. Razin. 1984. International capital movements under uncertainty. *Journal of Political Economy* 92 (2): 286–306.

Hodder, J. E., and L. W. Senbet. 1990. International capital structure theory. *Journal of Finance* 45 (5):1495–1516.

Krueger, D., and Fabrizio Perri. 2006. Does income inequality lead to consumption inequality? Evidence and theory. *Review of Economic Studies* 73 (1): 163–194.

Kumhof, M., and R. Rancière. 2010. Inequality, leverage and crises. IMF Working Paper 10/268 International Monetary Fund, Washington, D.C.

MacDougall, 1960. The benefits and costs op private investment from abroad: a theoretical approach. *Bulletin of the Oxford University Institute of Statistics,* Volume 22, Issue 3: 189–211.

Steiner, R. 1922. Nationalökonomischer Kurs, Vierzehn Vorträge, gehalten in Dornach in Juli bis August 1922, 5. neu durchgesehene und ergänzte Auflage, *Gesamptausgabe* Dornach 1979.

Turner, R. K., D. Pearce, and I. Bateman. 1994. *Environmental economics*. Aldershot: Harvester Wheatsheaf.

Van Schaik, A. 1996. The role of interest in the economic system, *master thesis Erasmus University*. Rotterdam.

Aloy Soppe is associate professor of financial ethics at the Erasmus University in Rotterdam. He has a Masters Degree in General Economics from the University of Groningen (RUG). He was a stock analyst for the Amsterdam Stock Exchange at the ABN/AMRO bank, and later worked as director of Soppe Currency Consultants and advisor for the implementation of option strategies in managing currency risk. Since 1987, he has worked at the Department of Finance and Investments at Erasmus University in Rotterdam. In 1993, he moved into sustainability research and started teaching Financial Ethics. In 2001, he joined the Erasmus University Law Faculty to teach Financial Ethics. He completed his PhD study entitled: Finance as an instrument to a sustainable company and publishes in national and international journals.

Spiritual-Based Entrepreneurship for an Alternative Food Culture: The Transformational Power of Navdanya

Arundhati Virmani and François Lépineux

This paper examines Navdanya's actions and proposals for spiritually-grounded sustainable agriculture in a national and international frame and analyses this organisation as an example of spiritual-based entrepreneurship with a multi-level transformational power. It first situates the emergence of Navdanya in the Indian national context, marked by the evolution of agricultural patterns and the spreading of entrepreneurial practice. Navdanya is a network of seed keepers and organic producers now covering 16 Indian states which advocates the practices of non-violent farming, sharing or gifting. It trains 'small' farmers and has established a learning centre, *Bija Vidyapeeth*—'Earth University'. The organization's growth and diversification in India now encompasses issues ranging from environmentalism to land sovereignty and is distinguished by its identification with spiritual notions specific to Indian political culture.

In the first half of the twentieth century, an ordinary, unknown lawyer, *Mohandas Karamchand Gandhi* demonstrated the power of spirituality in the political arena. Notions such as soul force or non-violence (*ahimsa*), words hitherto confined to religious movements and groups, came to constitute and impose a new vocabulary of political culture and practices of resistance. Today, still in India but in another century and in another sphere, a similar change is taking place with the introduction of spirituality as a frame of reference for agricultural practices. Navdanya mobilizes ideas that constitute an ideological frame for political action in India; they include the notion of non-violent agriculture, clearly inspired by Gandhi. These distinctively Indian characteristics and associated spiritual philosophy lend it national legitimacy and international recognition.

A. Virmani (✉)
EHESS, Centre Norbert Elias, Marseille, France
e-mail: arundhativirmani@gmail.com

A. Virmani · F. Lépineux
ESC Rennes School of Business, Rennes, France

F. Lépineux
e-mail: françois.lepineux@esc-rennes.com

L. Zsolnai (ed.), *The Spiritual Dimension of Business Ethics and Sustainability Management*, DOI 10.1007/978-3-319-11677-8_11

The second section of the paper concentrates on the peculiarity of Navdanya: its multi-faceted activities make it an unclassifiable actor—it is at the same time an NGO, a solidarity network, a militant co-operative movement, a private venture, a university and an environmental activist group with multiple international connections. The most appropriate conceptual framework for describing it is probably spiritual-based entrepreneurship. Navdanya offers an example of how spiritual values and traditions can be mobilized to propose a vision of sustainable agriculture, defend the individual against corporate rights and promote an alternative form of food culture. The multi-level transformational power of Navdanya should also be highlighted: it lays claim to being a catalyst for an economic model based on fairness, a lever for social change and an actor in defence of the common good. Its spirit can therefore be captured by the phrase 'transformational entrepreneurship', within the meaning that we will later specify.

1 The Emergence of Navdanya in the Indian Context

Changes in India's National Agricultural Policy

The Green Revolution of the 1970s, initially hailed as the harbinger of agricultural productivity, is today viewed more critically. It has led to a heavy reliance on fertilizers and pesticides and consequently to mono-cropping and a reduction in agricultural diversity. It is held to have caused soil salinization and degradation, with a spectacular increase in water consumption through public subsidies and abusive practices (Vyas 2003) and accelerated groundwater depletion even in rain-fed states (Andhra Pradesh, Karnataka, Rajasthan, Madhya Pradesh, Chhattisgarh and Gujarat). From 1990 onwards the thousands of farmer suicides that occurred in the cotton-growing districts of central and southern India were alternatively attributed to increased indebtedness (Gruère et al. 2008), to the introduction of Bt cotton, to cash cropping, industrial agriculture and globalization (Mahadevan 2003) as well as to developed-country subsidies that lowered world prices for cotton. The needs of India's growing population and the unstable trends of agricultural production coupled with the spate of farmer suicides gave urgency to the need to reform the mechanisms and regulatory institutions of rural farming.

The exclusive focus on food security—understood as a minimum quantity of available food –in the 1960s and 1970s in India gave way from the late 1980s onwards to broader questions about the environment, sustainability, common good and farmers' livelihoods. Movements that sought to defend small farmers and producers against large-scale development programs drew attention to the increasing number of farmers quitting cultivation activities: between the census of 1991 and 2001, nearly 8 million farmers quit farming. Given that agriculture and allied activities contributed nearly 50 % of India's national income and engaged around 72 % of its total working population, these trends raised questions about the type of national development strategies required to promote sustained agricultural growth in a rural landscape marked by a diversity of climate and crop patterns and practices (Ninan and Chandrashekar 1993). In particular, the expansion of sown areas became an important subject for debate (Tripathi and Prasad 2009). The idea that reductions in yield could be linked to land and water use and exploitation gained ground, and sustainable agriculture became a major concern in India. At the turn of the millennium, significant media coverage, both national and international, generated a flurry of reports and debates about agricultural and development policies (Krishnakumar 2005; Sahai et al. 2005; Nadal 2007).

The main challenges for Indian agriculture were now extended to minimizing environmental impacts or land degradation and preserving resources (land, water, air). India's National Agricultural Policy (NAP) placed sustainable agriculture and the importance of the management and conservation of resources at the heart of its focus: "the policy will seek to promote technically sound, economically viable, environmentally non-degrading, and socially acceptable use of country's natural resources—land, water and genetic endowment to promote sustainable development of agriculture" (Government of India 2000). Its highest priority was to improve the quality of land and soil, thereby sensitizing the farming community to environmental concerns and promoting more rational utilisation and conservation of water. The Tenth Five Year Plan for 2002–2007 emphasized natural resource management through rainwater harvesting and controlling groundwater exploitation, denounced the abusive use of fertilizers and recognized organic farming as a 'thrust area' (an area of focus) in the sustainable use and management of resources in agriculture (Government of India 2002).

This historical background explains the development of organic agriculture in India since the beginning of the millennium. Today, organic farming is seen as contributing to the protection of natural resources, to biodiversity and animal welfare and to the development of rural areas. It also stands for the greater economic stability of participating farmers and the opportunity to cultivate a wide range of produce (India currently exports organic products to Europe, the US and Japan). The Government of India launched the National Programme on Organic Production (NPOP) in 2000 to promote further development in this area. Navdanya's fight for rights and common good, articulated in a language rooted in the traditional wisdom and knowledge of common folk, fits into this context. Its contestation of WTO rules, of seed patents and of the invasive strategies of multinationals is strengthened by its counter proposals for sustainable agriculture. However, the shift from input-intensive to organic farming requires changes in a number of policy measures, from the transfer of information to training and the development of markets. Navdanya defines itself as a facilitator of these far-reaching changes.

Evolution of India's Entrepreneurial Landscape

Navdanya is a network of organic farmers and it is also an entrepreneurial venture; it is therefore interesting to trace its emergence in the context of changing entrepreneurship patterns in India. The practice of entrepreneurship in India has changed significantly since the beginning of the twentieth century. Early Indian entrepreneurs in the colonial period emerged from the ranks of general traders, indigenous bankers ('shroffs'), mill managers and technicians, merchants and financers. After independence, the planned economy system and its licence-raj (the exercise of industrial activities through government permits and state regulation of markets) froze business activities and entrepreneurship within family businesses and a traditional, paternalistic approach—in short, the age of the 'brown burra sahebs'. Government control of production and assistance in the form of funds and infrastructural facilities reduced the importance of entrepreneurial ability; the capacity to imagine and innovate or extend the frontiers of the market. What mattered was having the right contacts in government. The Nehruvian approach of the 1950s–1960s was sceptical of private enterprise and perceived it as primarily benefiting individuals, along with contradicting state development plans for the nation. Tata's famous description of his meeting with Nehru soon after independence, when Nehru doodled on a piece of paper while Tata tried to explain his vision of economic enterprise in a post independent nation, is a perfect example of this attitude.

Thus, until the 1990s, entrepreneurship and business leadership were predominantly developed and transmitted within family firms– big enterprises like the Tatas, Birlas, Thapar, Godrej, Bajaj, Mahindra, Goenkas, Khaitan or medium sized family enterprises—or by well-entrenched trading communities (Chettiars in the South, Marwaris in the East). The end of the protectionist era marked a rupture with this traditional pattern of entrepreneurship. A new model of entrepreneurship then gained ground, with the entry of people from diverse social groups and classes or occupational groups. Symptomatic of this kind of entrepreneurship was the figure of Dhirubai Ambani who gained entry into a select group of industrial houses whilst modestly claiming to be only a trader. The Ambani success story opened up new frontiers for self-made Indian entrepreneurs, heralding an era of the new entrepreneur, liberated from family traditions, caste approval, social roles and prestigious employment (civil service posts) and last but not least, community networks. The transformation of a small, unknown trader to a global presence did much to dissipate social prejudice against technical studies or industrial employment. The climate of entrepreneurship in India is considered more favourable today than half a century ago because there is less state intervention, better managerial skills and more focus on performance and turnover.

To sum up this brief historical review, sustainable agriculture has imposed itself as an important priority in India's development schemes over the last few decades, both for the government and for NGOs who act in defence of civil society. The growing interest in organic agriculture and the parallel evolution of entrepreneurial patterns have encouraged the emergence of a model of entrepreneurship that accentuates a holistic, spiritual approach to farming. It is in this context of renewal, and in an international climate that favoured and supported an alternative approach to agriculture, that Navdanya was born and then flourished; the 25th anniversary of the organisation was celebrated at the *Bhoomi 2012* festival of sustainable agriculture, seed sovereignty and womanhood.

Navdanya: A Movement Rooted in Indian Culture

The Navdanya movement grew out of a program of the Research Foundation for Science, Technology and Ecology (RFSTE) founded by the physicist and environmental activist Vandana Shiva in 1982. This Indian NGO defines itself as a network of seed keepers and organic producers and as a women-centred movement committed to supporting small farmers. It places rural producers, both men and women, in the forefront of its economic ventures and considers the idea of sustainable farming practices to be a viable marketing initiative; it has trained thousands of farmers in sustainable agriculture over the past two decades. In 1991, Navdanya was registered as a trust; since then it has expanded to become the largest direct marketing, fair trade organic network in the country, with a retail unit that sells products in stores in some of the major cities in India, mainly Delhi, Dehradun and Mumbai. This commercial activity enables small farmers, and especially women, to earn their livelihoods and provides them with the financial means to safeguard their indigenous knowledge and culture.

Navdanya's philosophy focuses on peace with nature through sustainable agriculture and organic farming, celebrating biodiversity—as symbiotic relationships between plants lead to higher yields on the same piece of land –, and solidarity economies rather than economies based on competition. It puts forward a holistic view of man's relation to the planet. It affirms the right to food as a fundamental right, raises awareness of the hazards of using chemical fertilizers and pesticides or genetically modified seeds, and combats bio-piracy and climate change. It defines its goals as organizing and promoting farmers' groups and preserving biological and cultural diversity, as well as battling for seed sovereignty, food sovereignty and water democracy—all themes covered by the slogan 'Earth Democracy'. Its university, *Bija Vidyapeeth*—'School of the Seed', or 'Earth University'—was founded in 2001 just after the 9/11 tragedy, on Navdanya's organic farm in Doon Valley, Uttarakhand. It promotes a vision of holistic solutions rooted in deep ecology as an alternative to the current world order dominated by greed, destruction and violence. In partnership with the Schumacher College in the UK it offers opportunities to explore the meaning of earth democracy, to learn from nature and to practice sustainability.

Among the distinctive features of Navdanya its spiritual and political dimension deserves to be noted. In opposition to ecologically questionable international agricultural politics, Navdanya mobilizes key elements of Indian political culture elaborated through Gandhian inspiration in the 1920s–1940s and merges them with global concerns of environment and planetary safety. Navdanya has successfully developed a spiritual language of resistance from a national anti-colonial tradition and battles to safeguard and revive Indian rural and artisanal activities. Its campaigns such as *Bij yatra* (seed pilgrimage), the *Jal swaraj* movement (protection of water) and *Bhu swaraj* (food security) recall earlier national battles for indigenous production and home industries and Gandhi's pilgrimage (salt *yatra*) to the sea to produce salt in defiance of British colonial laws. Its *Bija satyagraha* against the piracy of traditional knowledge and its control through patents by multinational, foreign firms, recalls Gandhi's non-violent *satyagrahas* against discriminatory legislation in South Africa and India. Navdanya's emphasis on *panchayats* (village assemblies) as bodies central to determining the use of common land also revives a Gandhian call to reactivate village democracy and strengthen the role of peasants in the economy. All these initiatives and stands uphold Gandhian notions of the struggle for truth and an even older nationalist vision of self-government that goes back to the early twentieth century; today they represent a call for a new freedom movement—the battle against Monsanto and WTO rules recalls the battle against the British government and colonial empires.

The very name *Navdanya* means 'nine seeds', suggesting the importance of protecting biodiversity. But it also means 'new gift' or 'new offerings': it contains the words *nav* (new) and *daan* or *dana* (gift, offering) and combines these two concepts, which are at the heart of many Indian traditions. *Nav*, new, is associated with renewal, rebirth, a fresh beginning. It holds forth promises and raises hopes. It is linked to auspiciousness, a key notion in Indian cultural tradition that is rooted in working culture. *Daan* (offering) is again a central principle that structures relations within the community, with kinsmen and with gods. Seed savers are also givers of seeds; their gift or *dana* is a gift of life and continuity. However, if these references belong to Indian traditions and Hinduism, they can also be found in other spiritual systems such as Christianity, Buddhism, or Native American religions.

2 Navdanya: The Transformational Power of a Multi-Faceted Movement

An Unclassifiable Actor

Navdanya's boundaries are difficult to pin down: it does not confine itself to the usual categories. Its multiple activities of protecting and training farmers, selling their products, fighting legal battles against bio-piracy and speaking up for the common good make it simultaneously a movement, an organisation, a protest group and an economic entity. Its national extent is impressive: it has implanted itself in 16 of the 28 Indian states, and more than 500,000 farmers with small and marginal farms (of less than 1 hectare) are primary members of Navdanya. It has set up 65 community seed banks across the country to preserve indigenous seeds and to offer an alternative to patent-controlled resources and production. Navdanya's seed banks conserve more than 2000 varieties of rice from all over India as well as a significant number of varieties of wheat, millet, vegetables and medicinal plants. Navdanya is at the same time an NGO, a solidarity network, a militant co-operative movement, a private venture, a university and an environmental activist group with multiple international connections. It distinguishes itself by its multi-pronged strategy and by a strong ideological basis rooted in India's anti-colonialist political culture. The following section of the paper analyses three distinctive facets of Navdanya: as an actor in the Indian organic market, an advocacy group, and as a movement with a global profile.

Actor in the Indian Organic Market

A major feature of Navdanya is its presence in the Indian organic market: in this respect Navdanya is an economic entity—it is a business which has a retail chain and a commercial strategy. It has a presence at all points in the supply chain, from production to distribution channels, sells over 100 organic products through its outlets in Delhi, Mumbai and Dehradun and aims to expand its network of outlets and cafés throughout the country. The variety of products sold includes rice, flour, spices, pulses, squashes, dry fruit, herbal tea, etc. The organic food market that Navdanya has entered is a small one, though its appeal is growing amongst the upper middle class and the number of firms in the field has risen. Among the major players of the Indian organic food market the following organisations can be mentioned:

- Conscious Food has pioneered organic farming and food in India since 1992, playing an important role in re-establishing ancient grains and encouraging farmers to use natural and organic farming methods to grow their crops; it sources its products from small organic farms and farming communities all over the country for its Mumbai factory.
- Organic India: a Lucknow-based producer and exporter of organic products founded in 1997. This business began as a small-scale purveyor of Ayurvedic herbal medicines but is now cultivating 50,000 acres of certified organic land; it markets its products throughout India and internationally (USA, Europe, the UK, Israel, Australia, Japan, etc.).
- 24 Letter Mantra, an organic food brand created in 2004 by Hyderabad-based Sresta, works with a network of over 10,000 farmers cultivating 30,000 acres across 11 states in India; it is committed to creating a healthy society (healthy farmers, healthy consumers) and preserving a healthy planet for a healthy tomorrow.
- Pristine Organics is present in specific areas of the nutrition sector such as malnutrition and lifestyle diseases, with ethically priced products in fields ranging from pediatric to geriatric care; it is concerned with preserving rare varieties of seeds and maintaining ecological equilibrium and its 'farm to table' approach is holistic in nature.
- Suminter India Organics focuses on six kinds of organic produce: fibre, oilseeds, cereals, spices, herbs and medicinal plants, and specialty products. This Mumbai-based company founded in 2003 has a triple bottom line approach in which it works to bring economic, social, and environmental benefits to the communities in which it has a presence.

Under current government policy, it takes approximately 3 years for a farm in conversion to be certified as organic. Farmers need financial assistance to make the transition from heavy state-subsidised inorganic farming to sustainable farming practices. To ensure that farmers grow organically, retailers who buy from farmers have to pay them a premium when yields are low as they may still be in the conversion period; this financial support is also useful as sales may be limited by the size of the market and the absence of preservatives or additives make organic products more prone to being damaged. Amongst the actors of the Indian organic market, Navdanya provides a participatory guarantee system of certification for farmers' products. Compared to other firms in the small but growing organic food market in India, Navdanya's approach, by placing marketing in the broader intellectual argument about the dysfunctionality of the current system, encourages qualitative changes in habits of production and consumption. Its commercial activity blends sales of organic products with the simultaneous constructive engagement of all points of the supply chain, working with citizens' movements, grassroots organizations, NGOs and governments in parallel. This differs from the commercial approaches of some (but not all) of its competitors who instrumentalize the terms and concepts 'organic' or 'spirituality' as more labels for use in effective and smarter marketing.

Advocacy Group

Since its inception Navdanya has led or participated in numerous campaigns in defence of the right to food and against bio-piracy or genetically modified seeds. It came into prominence through challenging U.S. companies' patents on *neem*—a tree commonly found in India whose seeds produce oil, used as a pesticide and fungicide. The *Neem Campaign* against bio-piracy started in 1994: Navdanya mobilized 100,000 signatures and started a legal process against the US Department of Agriculture and W.R. Grace patent on the fungicidal properties of *neem* in the European Patent Office at Munich. The patent on *neem* was first revoked in May 2000 and the EPO confirmed its decision in March 2005, revoking the patent in entirety, judging that there was "no inventive step" involved in the fungicide patent, thus confirming the "prior art" of the use of *neem*. In 1998, Navdanya started another campaign against a patent on Basmati rice involving the American company Rice-Tec. In August 2001 the United States Patent and Trademark Office revoked a large section of the patent, allowing for another victory by Navdanya against bio-piracy. In parallel, as Monsanto arrived in India in 1998, Navdanya started the *Monsanto Quit India* campaign on August 9th, the very day when Congress launched the *Quit India* movement against the British government in 1942.

Since the agreement on Trade-Related Intellectual Property Rights (TRIPs) came into existence as part of the international regulatory framework for the protection of Intellectual Property Rights (IPR), the IPR regime has been criticized for serving corporate interests and contributing to the fortunes of powerful corporations (Richards 2004) and worsening the economic dependence of poor countries (Matthews 2006). In particular, the TRIPs agreement has transformed natural products—seeds traditionally collected by farmers during every harvest—into private property liable to corporate patents, forcing rural communities to buy their seeds every season; this privatization has been denounced by many as 'global biopiracy' (Shiva 1997). Besides, it is clear that the uniformity of cultivated varieties may affect the gene pool and may render food security dependent upon the seed supply industry (Sreenath 2006).

Navdanya spearheads the growing peasant resistance movement against the global IPR regime and corporate convergence which bring about concentration and restrictions in the seed sector. In the 1990s Navdanya initiated the *Bija satyagraha* movement to resist the TRIPs agreement and subsequent IPR laws which grant corporations a monopoly on seeds and prohibit seed saving and sharing. In 1993, half a million farmers gathered in Bangalore for a *Bija Satyagraha* which constituted the first international protest against the World Trade Organization. The words Navdanya uses in its posters, 'ancestors', 'tradition', or *sankalp*, are drawn from the vocabulary of farmers and ordinary people. The Gandhian inspiration—civil disobedience, non-violence—appears clearly on the *Bija* poster which uses the image of Gandhi leading the salt march to the sea to produce salt in defiance of British colonial laws in 1931. Later on, Navdanya's *Bija Yatras* (seed pilgrimages) from Sewagram to Bangalore in 2006, from Champaran to Rajghat in 2009 and through Rajasthan in 2012 articulated civil society causes in a politically powerful language familiar to ordinary people, infused with spiritual goals and meaning.

Recently, Navdanya has taken up the cause of tribal rights against land appropriation by industries and private corporations. It opposes the forced displacement of tribals and sensitizes public opinion to the violation of their customary rights to land and forest by local governments and private firms. It has participated in a protest movement against the Vedanta Company's exploitation of *Niyamgiri*, a mountain held to be sacred for the *Dongria Kondh* tribe in Orissa. However, Navdanya is much more than a protest group and works along the lines (for example) of the *Chipko* movement of the 1970s that called for an end to deforestation (Vandana Shiva built close ties with Chipko in her younger years) (Astruc 2011). It is also more than an advocacy group, but is somewhat like *Gene Campaign*—a renowned research and advocacy organisation in Delhi which has been working to empower local communities to retain control over their genetic resources. Navdanya claims not only to advocate change but to make change happen. In particular, it regularly collaborates with many other actors—from civil society organizations to governmental bodies –, which has enabled it to make significant contributions to the Convention on Biological Diversity (CBD) and the Biosafety Protocol.[1]

Movement with a Global Profile

Though its commercial activity is restricted to the Indian domestic market, Navdanya has developed a significant ideological profile by linking up with different international movements for sustainability, for the defence of the South against the North, and for the championship of small producers, particularly women, against

[1] The *Cartagena Protocol on Biosafety* was adopted in January 2000 as a subsidiary agreement to the *UN Convention on Biological Diversity* and entered into force in September 2003. It is designed to regulate the international trade, handling and use of living modified organisms resulting from modern biotechnology that may have adverse effects on biological diversity, also taking into account risks to human health.

large multinational firms and a private sector whose major criterion remains profit. Navdanya's success as an internationally visible group is incontestable; its mobilization of historically significant and popular political concepts, through its firm anchoring in the Indian soil and national landscape, gives its actions in international platforms greater visibility and impact. The international outreach of Navdanya has recently gained strength: *Navdanya International* opened new offices in Florence—the announcement was made at *Terra Futura*, an annual event about living sustainably for an economically, socially and environmentally just world. This concluded a period of 10 years' collaboration between the regional government of Tuscany and the *International Commission on the Future of Food*, chaired by Vandana Shiva.

Navdanya's philosophy, goals and principles of action have drawn it into partnership with non-Indian movements like *Slow Food*—an international movement founded in 1989 that defends traditional regional cuisine and local economic systems that are committed to creating a future of food and agriculture in which small producers prosper, ethical consumption develops, and biodiversity and cultural diversity thrives (Grasseni 2012). Navdanya's stand, 'small is beautiful', converges with that of *Slow Food*: in 2001 Navdanya received the Slow Food Award and became a Slow Food partner—and Vandana Shiva has served as Vice-President of *Slow Food International*. Another important international partner of Navdanya is the *Gaia Foundation* which focuses on global issues of environmentalism and biodiversity and was founded in the mid-1980s by a group of ecological pioneers: Liz Hosken, an environmentalist from South Africa; José Lutzenberger, Brazil's prominent environmental activist; Wangari Maathai, founder of the *Green Belt* movement and recipient of the 2004 Nobel Peace Prize; Edward Posey, active in *The Business Network*—one of the first organisations to promote social and environmental responsibility in business—and Vandana Shiva. These pioneering environmentalists created the Gaia Foundation to raise public awareness about the dark underside of national and international development policies, mainly ecological degradation and social and economic inequity.

Navdanya's international partners also include *UBINIG* in Bangladesh, *Focus on the Global South*, *ETC Group* (Action Group on Erosion, Technology and Concentration) and many others. Several of these international movements share a common commitment to reviving cultural and biological diversity, regenerating healthy ecosystems, enhancing traditional knowledge and practices for land, seed, food and water sovereignty and strengthening community self-governance. The constitution of such an international network of partners has much to do with the charismatic figure of Vandana Shiva. Navdanya's strategy of being part of a global coalition bolsters its reputation and has proven to be more efficient when it comes to struggling against bio-piracy, influencing international debates about genetic engineering in agriculture and resisting mainstream globalization trends.

A Role Model for Spiritual-Based Entrepreneurship

As the above pages have demonstrated, Navdanya has shown its capacity to innovate; it has carved itself a niche in the competitive domestic organic market and has also set up an international platform. It has found creative responses to a situation in which small farmers fall victim to the big, global logic of multinational companies who would control genetically modified seeds and profit-oriented scientific research. Navdanya is an illustration of creative entrepreneurship that gives agricultural producers a new status and identity and re-infuses them with a national tradition: it gives them the responsibility for perpetuating traditional knowledge that is threatened with extinction. A relevant framework for seizing the nature of this movement is spiritual-based entrepreneurship.

Promoting a spiritual basis in agriculture and related activities is part of national policies in India. The 'Karuna Ratna' awards, instituted by Karuna International and the Acharya Hasti Centenary Celebrations Committee, recognise outstanding work on *ahimsa* (non-violence), love for ecology and vegetarianism; the awards were conferred in 2010 on animal welfare and environmental activists. Navdanya's model of spiritual entrepreneurship is grounded in the principles of peace and non-violence. Navdanya identifies its mission as the promotion of peace and harmony, justice and sustainability. It defines work and production as a deeper experience of responsibility and as the duty of actors towards society and their environment. 'Fairness' is among its keywords: fair trade, fairness to earth and her species, fairness to producers and to consumers. It underlines the importance of care and awareness for all the actors and elements in the production cycle, as well as the need to respect species, water, soil and diversity. It takes consumer groups to farms so that they can interact with member farmers in order to reinforce a collective commitment to sustainability and equality. In contrast to industrial farming which is criticised for having destroyed the relationship between nature and food, organic agriculture is upheld as a spiritual act, even as it is commercialized.

The use of the term 'true organic agriculture' is explained as acknowledging and maintaining the sacredness of life. Farming and agriculture from this perspective become a spiritual act. This theme is traced to the *Taitri Upanishad*. *Annadaana* (growing and giving of food in abundance) is considered to be the highest duty in the world; abundance cannot come from exploiting the soil or farmers. Navdanya maintains that the rehabilitation of Indian agriculture cannot be done by tinkering with the still dominant system. Much of the literature on sustainable agriculture deals with managing growth and promoting equity in harmony with nature, but still argues in terms of strategies of proper regulation and management, moisture and soil conservation and income-enhancing crops. Navdanya's spiritual entrepreneurship model, on the other hand, shifts the emphasis onto empowering small actors, recovering and preserving traditional knowledge and skills and defending common goods against the invading tentacles of intellectual property rights. The connection between humans, land and seeds is deeply embedded in cultural and historical contexts.

While Navdanya's spiritual dimension is rooted in the Hindu tradition, it is not limited to this particular tradition; it conveys a universal message, which finds echoes in Buddhism as well as Christianity and other religions. For instance, the emphasis placed by Navdanya on sustainable agriculture resonates with Buddhist economics as outlined by *Ernst Friedrich Schumacher* in his landmark work (Schumacher 1999) and more recently by *Sulak Sivaraksa*, a Thai intellectual and one of Asia's prominent thinkers (Sivaraksa 2009). Similarly, the aim of the common good pursued by Navdanya resonates with Christian ethics (Hollenbach 2003) and especially with the Catholic social doctrine (Naughton and Cornwall 2006).

Navdanya blurs the lines and pushes the boundaries of existing notions of entrepreneurship. Its philosophy of action and path of development not only hold the promise of a new pattern of socio-eco-development; they actively contribute to the coming into being of this alternative socio-eco-development model for agriculture and the food industry. Navdanya's strategy of global visibility gives its entrepreneurial activity a political role within India and on the global stage. It is, however, still too early to evaluate Navdanya's economic performance and come to definite conclusions about its pioneering role in outlining a spiritual, holistic approach to sustainable agriculture in the market. Certainly, it is a model of spiritual-based entrepreneurship that invites reflection and analysis.

Transformational Power and Transformational Entrepreneurship

The originality of Navdanya resides in its transformational power, which is multiform. Not only does the movement take the evolution of the Indian economy further, but it also acts to facilitate a shift in societal and cultural patterns and contributes to challenging the existing rules of globalization. Navdanya connects local initiatives, the national landscape and the planetary setting; it links the individual level to the global scale. Navdanya's transformational power simultaneously impacts individual farmers, local communities, the Indian agricultural sector, the economic system, the food culture and international regulations on IPR and agriculture.

Firstly, Navdanya acts as a catalyst for an alternative economic model; its economic philosophy is humanistic and closely tied to the thoughts of E. F. Schumacher as expressed in *Small Is Beautiful—Economics as if People Mattered* (Schumacher 1999). The educational programs of *Bijya Vidyapeeth*, the 'Earth University' created by Navdanya, are developed with the Schumacher College in the UK with a view to exploring and practicing the science and art of sustainability. Navdanya stands for the promotion of organic fair trade based on fairness to the earth and all her species, fairness to producers and fairness to consumers. By focusing on small-scale and indigenous solutions that foster personal development, it offers a sustainable alternative to the still-dominant economic model followed by agriculture and the food industry. It has a 'seed to table' and 'farmers to co-producers' programme which involves providing farmers with financial assistance, training them in sustainable agricultural practices and finding markets for their processed products. Navdanya takes its role even further and has set up 65 community seed banks that supply farmers with GMO-free seeds locally.

It may be argued that, as a solidarity network, Navdanya is taking part in the emergence of a 'sharing economy' (Gold 2004) which emphasizes cooperation before competition and promotes economic justice and democracy (Hahnel 2005). Navdanya fosters small-scale initiatives for local communities and pushes for systemic change on the national and global scales; its development constitutes an excellent illustration of the necessary quest for global justice at the beginning of the twenty-first century (Mitzman 2003). To borrow the title of Daly and Cobb's book, the Navdanya movement is committed to 'redirecting the economy toward community, the environment and a sustainable future' (Daly and Cobb 1994). This orientation echoes a philosophy that finds supporters at the global level: the United Nations conference *Happiness and Well-Being: Defining a New Economic Paradigm* convened by the Kingdom of Bhutan in April 2012 brought together numerous representatives from governments, religious organizations, academia and civil society to discuss this theme.

Navdanya's ability to engender social change encompasses the cultural dimension: the expansion of its network of outlets and cafes aims at creating another type of food culture in India—a culture which respects diversity, local production and food quality. Navdanya calls for a shift in values: food should not be a commodity, nor produced for profit but is life itself. It champions organic farming, organic fair-trade and organic food almost as a way of life: Vandana Shiva's conferences, articles and books repeat the centrality and importance of the message: organic food for everyone, not only for organic farmers. The emphasis is on bringing the sacred back into the food chain: the manner in which food is produced, processed, distributed and eaten must be sacralized if humanity has to have a future. To conserve seeds means to conserve biodiversity, knowledge and culture; moreover, sustainable agriculture may lead to the nurturing of creativity and culture. The societal influence of Navdanya resonates with current reflections on bioregionalism (Carr 2004), on the future of local cultures (Redner 2004), and on the construction of global civil society (Keane 2003; Chandler 2004; Glasius et al. 2006).

In the authors' view, Navdanya is an illustration of transformational entrepreneurship—within the meaning that is now specified. Transformational entrepreneurship as we see it shares some common traits with social entrepreneurship, insofar as many social entrepreneurs act as change agents in society, tackling social problems with new approaches that improve the lives of people living in poverty and triggering long-term societal effects. Social entrepreneurship is now well developed across the planet and a number of social entrepreneurs who pioneered problem-solving models can be identified (Bornstein 2007). To the extent that social entrepreneurship is a factor in societal transformation (Alvord et al. 2004), Navdanya can be considered a social entrepreneurship venture. However, it does not champion one single cause, or focus on the implementation of one single mechanism as other organizations who fall into this category usually do (for instance: combating illiteracy, bringing solar energy to villagers, providing health protection for children, etc.). It engages a chain of actors involved in production, sales and consumption; it challenges a wide range of institutional mechanisms—economic, legal and political; it leads or participates in multiple campaigns that push for systemic change. Therefore, we believe that the term 'transformational' is more adequate to depict the type of entrepreneurship incarnated by Navdanya.

In a recent paper Schoar (2010) drew a distinction between 'subsistence entrepreneurs' (who become entrepreneurs as this constitutes an opportunity for employment and a means of subsistence for them and their families), and 'transformational entrepreneurs' (who create medium or large-sized businesses that provide jobs for other workers and act as engines of growth in the economy). This distinction is quite relevant; however it limits the understanding of transformational entrepreneurship to its economic dimension whereas we envision it as a multidimensional phenomenon. We define transformational entrepreneurship as a form of entrepreneurship that paves the way for—and allows the realization of—the 'great transition' of the earth-system that lies ahead, and combines the three perspectives of foresight, global civil society and the common good. Transformational entrepreneurship fosters economic, societal and cultural change towards a global civilization that the planet can sustain and relies on the mobilization of three levers:

- Foresight—having a vision of the unfolding of the great transition ahead, of a desirable post-transition state for the world, and linking this to the present;
- Global civil society—raising awareness of the general public at all levels, from the local to the global, about the stakes of the great transition;
- The aim of the common good—emphasizing the importance of preserving and restoring nature, of promoting global justice, of placing the human dimension at the heart of economic relations.

Foresight

The current economic system is not sustainable as it provokes climate change, depletion of natural resources, biodiversity losses and the pollution of air, soil and water, among many other environmental consequences. The carrying capacity of the planet is not limitless yet the activities of mankind exceed by far its limits and 'overshoot day'[2] occurs earlier every year (the date for 2012 was Wednesday, August 22)[3]. Moreover, globalization has ambivalent social effects; if it is true that hundreds of millions of people have been lifted out of extreme poverty, the so-called 'bottom of the pyramid' is still crowded and inequalities have risen dramatically over the past two decades. The poorest are still deprived of fundamental human rights, of access to primary education, of drinkable water, of health services, of basic foods. The price of oil, of cereals, of metals and of rare earth materials is on the rise; these trends, as well as the costs generated by extreme meteorological events, are likely to destabilize the global economic system.

For these reasons and many others, a 'great transition' towards a sustainable and humane civilization is both desirable and inevitable if runaway reactions and global chaos are to be avoided. This transition will be—and already is—complex and multidimensional: it concerns the economic and financial system, earth's climate, energy, resources, biodiversity, human demography, society, culture, public health,

[2] The day when humanity starts using more natural resources than the earth can produce in 1 year in a sustainable way.

[3] In 2011, 'overshoot day' was September 27.

spirituality—and this brief list is not exhaustive. Transformational entrepreneurship is a form of entrepreneurship which triggers, accompanies or accelerates the unfolding of the great transition that is ahead of us—the transition of the world as a whole towards an economy that respects nature, towards a sustainable society, towards a humane civilization.

Global Civil Society

Transformational entrepreneurship involves the entrepreneur's commitment to raising the awareness of the general public about the forthcoming transition through a variety of means in the country where they operate, as well as beyond national borders. Navdanya raises awareness of civil society through several means and at all levels, from the local to the global. It also creates a discursive space to reinforce the movement through reflection on modes of action. It has organized extensive protest and advocacy campaigns, led civic mobilizations, developed women's empowerment programs and participated in slow food festivals. It provides alternative education at the Doon Valley organic farm and conducts programs for children in schools (e.g. through setting up organic farming plots). Navdanya is involved in seminars, workshops and events throughout India with farmers, local partner organizations or as a part of active social movements.

The influence of Navdanya extends well beyond the borders of India: it is also present in international fora and institutions and therefore has a global outreach. Vandana Shiva disseminates the organization's message around the globe as an invited lecturer and through films and books that target a wide audience. For example, the *Bhoomi 2012* festival included a conference entitled *Women in Defence of the Earth* which was held at the India International Center in Delhi on October 1st, bringing together leading eco-feminists including Frances Moore Lappé (author of *Ecomind*), Radha Bhatt (Chipko), and Vandana Shiva. The festival marked the launch of Navdanya's new campaign, the *Global Campaign on Seed Freedom* which aims to stop seed laws that prevent farmers from freely saving, distributing and exchanging native varieties of seed.

The Aim of Striving for the Common Good

Transformational entrepreneurship is thus defined with reference to its finality—the search for the common good; it has a teleological component. At a time when the situation of ecosystems and biodiversity worldwide is a major cause of concern (Millennium Ecosystem Assessment Report 2005) and prospects regarding climate change are bleak (Intergovernmental Panel on Climate Change 2007), the twin problems of providing global public goods (Kaul et al. 2003) and governing the commons (Ostrom 2008) are at the centre of attention for an increasing number of politicians, researchers, NGOs and civil society organisations. Globalization calls for a new, global ethic (Küng 2004); the ideal of the common good—or its equivalents in various cultures—should be one of the overarching principles on which a world community is built (Baudot 2001). Navdanya envisions sustainability as being within the broader perspective of the common *good* and views indigenous seeds as common *goods*. All farmers should have the right to save and share seeds; organic farming and sustainable agriculture are possible only if biodiversity is free from the threats posed by genetic engineering and seed patents.

Navdanya's battles include restoring common land to local communities and preventing the transformation of productive village commons into industrial plantations. It is at the head of popular resistance movements for the protection of the commons (such as *Mannu Akshana Koota,* or 'Movement for Saving the Soil'). It leads protests against the 'Tree Patta Scheme' funded by the World Bank, which promotes deforestation by leasing common lands to individuals and enabling planting through loans. Repayment of these loans creates pressure to produce marketable products, not sustainable forestry. Navdanya's concern for the common good found interesting echoes at the Zermatt Summit held in June 2012, which issued a 'Declaration on the Common Good to Humanise Globalisation'.

3 Conclusion

Navdanya is undoubtedly an extremely rich object of study. It defines itself as a network of partners, and as such it promotes solidarity within and between farmer groups, rural villages and local communities. It champions organic farming, encourages qualitative changes in production and consumption modes, and aims to create another type of food culture in India by addressing itself equally to school children and young adults. It calls for a shift in values and for a new freedom movement. The principles of self-reliance and the independence of farmers over their lives lie at the roots of the organization, and the choice to maintain crop diversity may be understood as a political stance. Navdanya has consistently stood up for bio-safety and warned against the dangers of genetically modified organisms. It actively promotes an alternative economic model and fosters social change; its expansion is part of a growing international movement of support for local cultures against unfair global rules. It links up with different international movements for sustainability, for the defence of the South against the North, and it champions small producers, especially women. It can certainly serve as an example for similar grassroots initiatives elsewhere—in the same sphere of action or in a different one.

From the point of view of organisational science, Navdanya's status is unclear: it is at once a solidarity network, an NGO, an economic venture, a cultural movement and an influential force that challenges the rules of globalization. It may be considered a *frontier object* that blurs the lines. Navdanya has a transformational strategy and acts as a catalyst for change in the economic, societal, cultural and political arenas, both in India and internationally. Its development infers the emergence of a new form of entrepreneurship: *transformational entrepreneurship*. As Navdanya plays a pioneering role in the evolution of agriculture in India and beyond, the question of whether the organisation marks a new stage in the evolution of entrepreneurship in India deserves attention. Moreover, the breadth and depth of the transformational nature of this movement require further investigation. Finally, the concept of transformational entrepreneurship as we see it needs to be more robustly elaborated—but this will be the subject of another contribution.

References

Alvord, S.H., D.L. Brown, and C.W. Letts. 2004. Social entrepreneurship and societal transformation: An exploratory study. *Journal of Applied Behavioural Science* 40 (3): 260–282.

Astruc, L. 2011. *Vandana Shiva. Victoires d'une Indienne contre le pillage de la biodiversité*. Mens: Terre vivante.

Baudot, J., ed. 2001. *Building a world community: Globalisation and the common good*. Washington, DC: University of Washington Press.

Bornstein, D. 2007. *How to change the world: Social entrepreneurs and the power of new ideas*. New York: Oxford University Press.

Carr, M. 2004. *Bioregionalism and civil society. Democratic challenges to corporate globalism*. Vancouver: UBC Press.

Chandler, D. 2004. *Constructing global civil society*. London: Palgrave Macmillan.

Daly, H., and J. Cobb. 1994. *For the common good. Redirecting the economy toward community, the environment and a sustainable future*. Boston: Beacon Press.

Glasius, M., M. Kaldor, and H. Anheier, eds. 2006. *Global civil society*. London: Sage.

Gold, L. 2004. *The sharing economy. Solidarity networks transforming globalization*. Aldershot: Ashgate.

Government of India. 2000. *National agriculture policy—2000*. New Delhi: Ministry of Agriculture.

Government of India. 2002. *Tenth five year plan 2002–2007*. Vol. II. New Delhi: Planning Commission.

Grasseni, C. 2012. Re-inventing food: The ethics of developing local food. In *Ethical Consumption: Social Value and Economic Practice*, ed. J. G. Carier and P. G. Luetchford. Oxford: Berghahn Books.

Gruère, G.P., P. Mehta-Bhatt, and D. Sengupta. 2008. Bt Cotton and Farmer Suicides in India. Reviewing the Evidence. *IFPRI Discussion Paper n°808*, International Food Policy Research Institute, Washington, DC.

Hahnel, R. 2005. *Economic justice and democracy. From competition to cooperation*. New York: Routledge.

Hollenbach, D. 2003. *The common good and Christian ethics*. Cambridge: Cambridge University Press.

Intergovernmental Panel on Climate Change. 2007. *Fourth assessment report*. Geneva: IPCC.

Kaul, I., P. Conceiçao, K. Le Goulven, and R.U. Mendoza, eds. 2003. *Providing global public goods: Managing globalization*. New York: Oxford University Press.

Keane, J. 2003. *Global civil society?* Cambridge: Cambridge University Press.

Krishnakumar, A. 2005. Seeds of controversy. *Frontline* 22 (12): 4–17.

Küng, H. 2004. *Global responsibility: In search of a new world ethic*. Eugene: Wipf & Stock Publishers.

Mahadevan, R. 2003. Productivity growth in Indian agriculture: The role of globalization and economic reform. *Asia-Pacific Development Journal* 10 (1): 57–71.

Matthews, D. 2006. *Globalising intellectual property rights: The TRIPS agreement*. London: Routledge.

Mitzman, A. 2003. *Prometheus revisited. The quest for global justice in the twenty-first century*. Boston: University of Massachusetts Press.

Nadal, A. 2007, Jan. 6. Monsanto, Cereal Killer GM and agrarian suicides in India. *La Jornada*.

Naughton, M.J., and J.R. Cornwall. 2006. The virtue of courage in entrepreneurship: Engaging the catholic social tradition and the life-cycle of the business. *Business Ethics Quarterly* 16 (1): 69–93.
Ninan, K.N., and H. Chandrashekar. 1993. Green evolution, dryland agriculture and sustainability: Insights from India. *Economic and Political Weekly* 28 (12): A2–A7.
Ostrom, E. 2008. *Governing the commons: The evolution of institutions for collective action*. New York: Cambridge University Press.
Redner, H. 2004. *Conserving cultures. Technology, globalization and the future of local cultures*. Lanham: Rowman & Littlefield Publishers.
Richards, D.G. 2004. *Intellectual property rights and global capitalism: The political economy of the trips agreement*. New York: M. E. Sharpe.
Sahai, S., B. Mahanta, and W. Ahmed. 2005. *Farmer's perception of agrobiodiversity: A study in North India*. New Delhi: Gene Campaign.
Schoar, A. 2010. The divide between subsistence and transformational entrepreneurship. In *Innovation Policy and the Economy*, ed. J. Lerner and S. Stern, 57–81. Chicago: University of Chicago Press.
Schumacher, E.F. 1999. *Small is beautiful. Economics as if people mattered*. Point Roberts: Hartley & Marks Publishers. (First published in 1973).
Shiva, V. 1997. *Biopiracy: The plunder of nature and knowledge*. Cambridge: South End Press.
Sivaraksa, S. 2009. *The wisdom of sustainability. Buddhist economics for the 21st century*. London: Souvenir Press.
Sreenath, L. 2006. Impact of post-trips I.P. Regime on Indian agriculture. *CA Journal* 54 (8): 1178–1185.
Tripathi, A., and A.R. Prasad. 2009. Agricultural development in India since independence: A study on progress, performance, and determinants. *Journal of Emerging Knowledge on Emerging Markets* 1 (1): 63–92.
United Nations. 2005. *Millennium ecosystem assessment report*. New York: UN.
Vyas, V.S. 2003. *India's agrarian structure, economic policies and sustainable development*. New Delhi: Academic Foundation Publishers.

Arundhati Virmani did her PhD from the Sorbonne, Paris. She taught in Delhi University, University of Bordeaux and currently teaches at the EHESS, Centre Norbert Elias, Marseille. She is a business consultant in intercultural relations. Her research focus is colonial and contemporary India, particularly recent political, cultural and religious transformations in Indian society. Her books include "L'Inde, une puissance en mutation" (Paris, La Documentation Française, 2001), "India: 1900–1947" (Paris, Autrement, 2001), "A National Flag for India. Rituals, Nationalism and the Politics of Sentiment" (Delhi, Permanent Black, 2008) and "Atlas de l'histoire de l'Inde" (Paris, Autrement, forthcoming, 2012)

François Lépineux is professor at ESC Rennes School of Business (Brittany, France), where he is in charge of the CANON Chair "Foresight and the Common Good" and Head of the Centre for Responsible Business. After graduating from HEC Paris School of Management in 1990, he has carried on various research and consulting activities. He received his PhD in Management Science at the Conservatoire National des Arts et Métiers (CNAM, Paris). He has co-edited two books with Professor Henri-Claude de Bettignies: "Finance for a Better World: The Shift Toward Sustainability" (2009. London, Palgrave Macmillan) and "Business, Globalization and the Common Good" (2009. Oxford, Peter Lang Academic Publishers). He is co-author of the first handbook on CSR in French.

The Enforcement of the Self-Interests of Nature Transformers

Janos Vargha

The following story, almost 300 years old, was written by Jonathan Swift and published in his famous satirical novel, *Gulliver's Travels*. Gulliver tells us about his visit to Lagado:

> His lordship [...] only desired me to observe a ruined building, upon the side of a mountain about three miles distant, of which he gave me this account: That he had a very convenient mill within half a mile of his house, turned by a current from a large river, and sufficient for his own family, as well as a great number of his tenants; that about seven years ago, a club of those projectors came to him with proposals to destroy this mill, and build another on the side of that mountain, on the long ridge whereof a long canal must be cut, for a repository of water, to be conveyed up by pipes and engines to supply the mill, because the wind and air upon a height agitated the water, and thereby made it fitter for motion, and because the water, descending down a declivity, would turn the mill with half the current of a river whose course is more upon a level. He said, that being then not very well with the court, and pressed by many of his friends, he complied with the proposal; and after employing a hundred men for two years, the work miscarried, the projectors went off, laying the blame entirely upon him, railing at him ever since, and putting others upon the same experiment, with equal assurance of success, as well as equal disappointment (Swift 1726).

In this paper I will show that although Swift describes a fictitious project, this satirical story reflects the reality of many water construction megaprojects. There are numerous tangible examples of similar ambitious projects that have transformed the landscape and the environment—profitable for the projectors but disastrous for others, built because of similar arguments, employing similar methods.

J. Vargha (✉)
Budapest, Hungary
e-mail: janos.vargha@gmail.com

L. Zsolnai (ed.), *The Spiritual Dimension of Business Ethics and Sustainability Management*, DOI 10.1007/978-3-319-11677-8_12

1 Symbolic Interpretations

Several meanings are attributed to the quoted episode, especially to the ruined mill (Rogers 1975; Reilly 1982; Higgins 1994; Descargues-Grant 2011). Such interpretations are quite diverse: the metaphor may relate to a contemporary financial scandal (the bursting of the South Sea Company Bubble in 1720); tinkering with the social structure by foolish and corrupt meddlers; the illusion of progress or perhaps an attempt to replace one dynasty with another. Beyond these symbolic interpretations, the news about numerous engineering projects of that time (many of them quite peculiar and/or fraudulent) might have served as inspiration for Gulliver's Travels. Pat Rogers gives a lengthy list of such projects, many of which are related to water engineering (Rogers 1975).

The Absurdity of Reality

One point in the story told by Gulliver, when the projectors argue about the agitation of the water by the wind and of air at height, seems to be so absurd that one might believe that it is a satirical technique to make the targeted person(s) look foolish. However, examples of similarly absurd argumentation, no less ridiculous, can be found in the real world. Elemér Sajó, hydraulic engineer and leader of the Hungarian civil engineering service in the Ministry of Agriculture from 1930 until 1934, wrote in a memorandum urging that lowland canals and reservoirs be built, that "aviation is safer over the waters and plains than mountains" (Sajó 1931).

The arsenal of the Swiftian Club of Projectors consists of *exaggerated* or *false promises*, *political pressure*, and *scapegoating*. This list can be extended with more items by examining the activity of projectors from antiquity to current times. Projectors, not only Swift's contemporaries but many others during the previous millennia and the following centuries, frequently ignore negative experiences, or at least do not learn from their mistakes. Accordingly, they make the same errors again and again.

History is littered with infamous landmarks left behind by the Club of Projectors: ancient irrigation canals in Mesopotamia that salinified soils and made them infertile; the Corinth Canal in Greece, one of Nero's projects that was completed in the nineteenth century but never yielded the promised benefits; the Grand Canal of Alsace that was abandoned at half its planned length because of negative impacts (but the part which has been built, from which most of the water of the Rhine is derived, caused the groundwater table to drop significantly and large wetland forests to perish); the Aral sea that has almost completely dried out as a consequence of the construction of irrigation canals that deprive its two feeder rivers of water. These are just a few examples from a very long series.

Hydropolitics

A common and distinctive feature of the majority of such projects is that they are linked to the state and to politics. One main reason for this is that, over a certain size—and many of these types of water projects belong to the big, mega or giga category –, their building takes a long time, not infrequently decades, and locks up huge financial resources for long periods of time with very uncertain returns. Another main reason is that the projects usually occupy, influence or reduce in value private and/or state properties, provoking stakeholders to actively or passively resist them. In addition, supporters of hydropower projects must strongly compete against other projectors who offer other types of power plants which are cheaper and require less time for construction. Recently, supporters of such projects have also encountered public opposition in the form of environmental groups as more and more knowledge accumulates about the negative ecological impacts of such projects. Promoters, therefore, are "desperately looking for justifications for public subsidies" (McCully 2001), and are seeking support from any kind of political systems (Vargha 1997).

2 Problem Solving by Projectors

How do the promoters of such projects manage these and other kind of conflicts that hinder the achievement of their goals? Here I will recall some examples, partly from the history of the Club of Projectors and partly using the case of the Gabčíkovo-Nagymaros hydropower plant.

Promise success against rivals to politicians, political parties or even to the nation. In the first year of World War I, Ödön Bogdánfy, hydraulic engineer, tried to exploit the excited sentiments of Hungarians against the other nations of the Austro–Hungarian Monarchy: "…in addition to general economic interests, hydropower utilization has a special national interest. With help of new plants constructed in the [mountainous] regions inhabited by other nationalities we will establish new centers for the Hungarian people, which will have outstanding influence in assimilating those nationalities" (Bogdánfy 1914).

Elemér Sajó outlined how channelized rivers and new canals could be used against the social unrest of the proletariat. "Factories are unwanted around political centers as the artificially heated discontent and demonstrations of the concentrated working masses may disturb the workings of administration and legislation". The factories—suggests Sajó—"should be dispersed alongside the banks of navigable rivers and canals, where the workers can nurse their children in the clean, healthy air of the villages, not in the slums of cities" (Sajó 1931).

The first attempt at the end of the nineteenth century to build the Grand Canal of Alsace by replacing the original bed of the Rhine with a 120 km long insulated lateral canal between Basel and Strasbourg failed as the Baden-Württemberg government (which had jurisdiction over both banks of the river) did not accept the plan to abandon the riverbed. The main reason for the refusal was concern about potential negative impacts on agriculture. During World War I France occupied Alsace and the Rhine became the border with Germany. The promoters renewed the idea of building the canal by arguing for the political and military advantages of having the waterway in French hands (in this way France could control both banks of the waterway and the Rhineland). French politicians considered the proposal and the Allied Powers at the Paris Peace Conference granted the necessary permission for France to include the project in the terms of the Versailles Peace Treaty. However, the project was never completed after several unsuccessful attempts to mitigate negative impacts on water supply, urban sanitation, agriculture and nature (Vargha 1984; Chioc 2002).

At the peace negotiations in 1919, following the example of the Rhine, the newly-established Czechoslovakia attempted—without success—to obtain similar exclusive rights to hydropower on the Danube River sections that constitute the border between Austria and Hungary. This was the very first step in a long process that culminated in 1992 in the unilateral diversion of the Danube River into a canal very similar to the Grand Canal of Alsace. Due to the protracted, combined efforts of (Czech-)Slovak and Hungarian project promoters and developers Slovakia finally received the right to the project in the form of a bilateral treaty on the construction of the Gabčíkovo-Nagymaros project, signed in 1978. By this time France had already relinquished her exclusive rights to (and had abandoned trying to complete) the Grand Canal of Alsace (Vargha 1984).

Make the argumentation fit the political ideology. In Hungary between the two World Wars German hydropolitics was the exemplar to be followed. "The waterway of the Danube played a huge role in the war efforts of the Central Powers. The perfectly equipped German waterways [...] played an extraordinary role in the amazing, tough and long-lasting resistance of the Central Powers against a superior force. Even during the War the Germans recognized that if they lost their overseas colonies, they could work off eastward the population surplus that was growing year by year with its resulting overflow of industrial vigor. They started to recognize, even at that time, that in addition to the Rhine-Danube canal they need a waterway that lead directly toward Russia" (Sajó 1931). This rhetoric was in harmony with the pro-German orientation of the Horthy regime. The intention was inevitably to obtain political support and public subsidies for the canals, reservoirs and dams that Sajó urged the government construct.

After World War II, a huge river channelization and damming scheme and its organizational body, the Tennessee Valley Authority in the United States, was presented as an example for constructing hydroelectric power plants on the Danube (Kiss 1947).

That decisive year, 1948, bought a turning point in the political orientation of Hungary and in the rhetoric of the projectors. Communists seized power and engineers started to refer positively to the hydropower constructions of the Soviet Union. In a lecture at the Hungarian Academy of Sciences in 1950, Emil Mosonyi, a leading Hungarian water engineer, said with great solemnity: "The *Soviet Union's* recently-published *gigantic plans* sharply illuminate the extraordinary importance of hydropower management.[1] The extraordinary size and capacity of the Kuybyshev and Stalingrad hydropower stations that were planned for the Volga River, and moreover the number and magnitude of challenges that are to be solved during the implementation of these projects suggest that our technical conceptions cannot remain within their old, usual, narrow-minded limits. [...] The news about those monumental technical creations should also influence us. Let us stand for a while in front of the map of Hungary in a festive mood with the desire to construct, generated in this way! Let us investigate domestic potential and sum up the tasks ahead of us!" (Mosonyi 1951).

After the fall of the Stalinist Rákosi era and under the more pragmatic but still totalitarian Kádár-regime, ideological argumentation became less extreme. Gyula Jolánkai, a civil engineer who promoted the construction of dams and reservoirs was already referring to examples of the "gigantic water projects" of the United States and Canada, as well as to those of the Soviet Union, and also to projects underway in neighboring countries. Jolánkai—in an attempt to tug at the reader's heartstrings—mentioned that, in this respect, "we are among the last ones", and that "in spite of the Tiszalök barrage, we lack large, representative water projects" [which are exclusively suitable for showing the economic role and importance of water management], as other, smaller constructions are "hidden under the roads of cities" or are "far from settled areas" and are thus "absolutely unsuitable for awakening public attention" (Gyula Jolánkai 1957).

Recently, remarkable signs of rapprochement between ideological argumentation have occurred. A Hungarian 'anti-environmental' group of projectors, the 'Real Greens'[2] clamors for the completion of the Gabčíkovo-Nagymaros project. They justify this call by assessing the present and future political situation in Hungary and predicting that the extreme right will help to achieve this goal. On the front page of their website the group declares in Hungarian and German [!] that Jobbik, a Hungarian neo-Nazi party, is the "first parliamentary party whose opinion about water management is based on centuries-old Hungarian traditions and prevailing world trends".[3] Based on this political orientation the group have attacked leaders of the Danube Circle, writing that "only second and third generation holocaust survivors are suitable for disseminating falsification and fraud".[4]

Associate water projects with national pride. The Gabčíkovo-Nagymaros project was deemed an object of national pride when Slovakia divorced from Bohemia. In the election campaign of 1992, shortly before the unilateral closing of the Danube by Czechoslovakia, the project was regarded in a political sense as "a unifying symbol of all Slovaks" (Fitzmaurice 1996).

[1] Emphasis in the original.

[2] The group has member and honorary members who are also members of the Hungarian Academy of Sciences. http://realzoldek.hu/external.htm. Last accessed October 6, 2013.

[3] http://realzoldek.hu. Last accessed December 22, 2013. The declaration in German: "Die Rechte ist die erste parlamentarische Partei, deren Meinung über unsere Wasserwirtschaft auf den seit einem Jahrhundert bestehenden ungarischen Traditionen und den in der Welt vorherrschenden Tendenzen beruht".

[4] http://www.realzoldek.hu/modules.php?name=News&file=article&sid=3019. Last accessed December 22, 2013.

Opponents of the project thus extolled were accused of trying to undermine the whole Slovak nation. Július Binder, general director of the Slovak water construction company Vodohospodárska Výstavba and a member of the Slovak parliament, claimed in an interview that the…

> (Danube Circle's) aim was not (and is not) to search for the best environmental and economic solution, but rather to stop the completion of the dam, to damage the Czechoslovak—and especially the Slovak—economy, and to exacerbate the controversy between both Slovaks and Hungarians and Slovaks and Czechs. In this unstable political and economic situation they could better pursue the policy of autonomy for Hungarian minorities and, later, the restoration of the Hungarian Empire. (Williams 1995)

Slovakia had to cover the construction and maintenance costs of the project from a foreign loan but the resulting income from electricity production was not enough to pay them back. More than a decade later, Gabčíkovo was still "a major economic burden for Slovakia" but this fact remained "largely hidden from the public and the dam is still presented as an object of national pride" (Zinke 2005).

Making gains from nationalism is a favorite tool of projectors and "such symbolism [is] displayed through megaprojects throughout the world" (Lunstrum 2011). For example, the GAP, a Southeast Anatolian project with 22 dams and 17 hydroelectricity plants in Turkey "has become a symbol of national pride, which seems to receive not only unanimous support from political parties across all ideological orientations but also maintained its privileged position within government's budgets" (Carkoglu and Eder 2001).

Cheat the common weal. In English satires of the sixteenth century, written even before Swift's work, the projectors always presented their projects "as progress for the public good" (Ratcliff 2012). Politicians who back projects that are advertised as promoting progress and public welfare may win votes as they are seen to be great friends of the public. The construction of many hydroprojects in the United States and elsewhere have been supported in this way; this style of politicking is sometimes called *pork-barrel politics*[5] (Goldsmith and Hildyard 1984; McCully 2001).

Projectors have a great propensity to claim that their dam projects are *multipurpose* or *complex* and will foster progress and public welfare in many fields (energy production, navigation, agriculture, water supply, recreation and tourism, etc.). In the early planning phase of the Gabčíkovo-Nagymaros project, even the projectors themselves stated that the barrages planned for the Danube between Bratislava and Budapest were nothing but hydropower plants, and therefore "the construction costs of dams and navigation locks on the Danube cannot be allocated to other water management interests, not even in a forced way" (Mosonyi 1948). However, on the section of the river designated for the Gabčíkovo-Nagymaros project investment costs were enormously high, with moderate energy-producing capacity compared to other electricity-generating technologies with much shorter construction times and lower costs per kilowatt hour. To overcome these drawbacks the projectors had two major tactics. First, they redesigned the projects several times to screw out as much

[5] The phrase apparently derives "from the frenzy of the hungry slaves on Southern plantations when their owners marked an occasion by rolling out a barrel of salted pork" (McCully 2001).

electricity from the river as possible; secondly, they arbitrarily modified the project budget to allocate as much cost as possible to other sectors (Vargha 1981).

The first tactic resulted in the implementation of a monstrous megaproject with only a few percent greater electric capacity than originally planned, but with far greater negative environmental impacts. The application of the second tactic resulted a significantly doctored budget, whereby the cost of the power plant itself (originally budgeted at 90 % of total project cost) was made equal to the cost of an equivalent thermal power plant (60 %). The remaining costs were loaded onto navigation (16 %), water management (11 %) and infrastructure (13 %) (Vargha 1981). From this perspective the projectors could advertise the Gabčíkovo-Nagymaros project as being a *complex utilization* of the Danube. However, this was rather an overstatement. As Parasuraman and Sengupta (2001) note, "often multi-purpose dams experience[d] greater shortfalls than single purpose dams due to over-optimistic targets".

Exploitation of forced laborers is another morally problematic way of decreasing a project's construction costs. In ancient times, 6000 Jewish prisoners were used in the unsuccessful attempt to dig the Corinth canal. In the nineteenth century, Egypt subsidized the building of the Suez Canal with "up to 30 million francs in the form of the value of the forced labor of the corveé system" (Weijnen 2010). In the Nazi era thousands of Jews worked on the construction of dams and power plants in Germany and Austria (Gruner 2006), and the forced laborers of Stalin's Soviet Union were used to build huge hydropower plants (Gregory and Lazarev 2003). In Romania prisoners of labor camps worked on the construction of the Danube-Black Sea Canal (Rothschild and Wingfield 2000). In Hungary, the Tiszalök dam on the Tisza River was built by political prisoners of the Rákosi era (Görbedi 1989; Rainer 2002). The chief engineer of the construction activities at Tiszalök was Emil Mosonyi, one of the most ambitious Hungarian projectors. In 1953, when Imre Nagy became prime minister, the use of labor camps was suspended and prisoners were released—with one exception. In Tiszalök the projectors maintained the camp in order to ensure that construction continued using the cheap manpower. The result was a protest by the severely disappointed prisoners. Guards from the State Protection Authority shot five of them (Görbedi 1989). A year later, at the inauguration of the Tiszalök dam, the chief projector, Emil Mosonyi was decorated with the Order of the Red Banner of Labour (Vargha 1981).

Override reasonable engineering limits. The above-mentioned method of slightly increasing the performance of the energy infrastructure lead to a conspicuously extreme scheme in the case of the Gabčíkovo-Nagymaros project. Reasonable limits (that were respected by the projectors themselves at the beginning) were later overstepped (Vargha 1981). These redesign steps were:

1. constructing one higher dam in the side canal instead of two lower ones;
2. adding a big, 60 km^2 lowland reservoir to the scheme;
3. doubling the the Gabčíkovo power plant's capacity to operate it at daily peak load;
4. decreasing water flow in the abandoned river bed to an extreme minimum;
5. extensively dredging the river bed downstream of Nagymaros.

In 1953, despite unambiguous opinions to the contrary, Mosonyi and his colleagues promised the Czechoslovak projectors that they would obtain consent of the Hungarian government to divert the river to Czechoslovakia *within 2 weeks*. This assurance was without result; the Hungarian deputy prime minister Ernő Gerő rebuffed the projectors' rash proposal. He declared that such a decision could not only be based on technical considerations but should be combined with an investigation of general economic impacts, including those of the dam on agriculture, irrigation, etc.—and also after addressing political concerns. Gerő expressed his dissatisfaction that the projectors were not at all able to give an acceptable answer to questions that arose about the likely drop in the groundwater table (estimated to be roughly 4–5 m) (Record of Meeting 1985).[6] The treatment of this issue at that time and the following decades well-characterizes how projectors act. The fall in the groundwater table is a direct consequence of diverting the river water into an artificial lateral canal (insulated with concrete and asphalt), and its height and existence in the highly permeable subsoil of the affected area depends essentially on any fall in the water level in the original river bed. However, the amount of electric energy (the project's main marketable product) that is produced depends on the flow of water to the turbines of the power plant that is built into the dam at the end of the upstream section of the lateral canal. From the projectors' point of view, any water left in the original river channel represents a loss. At the outset, without any investigation into or knowledge about potential impact, the projectors suggested leaving 200 m^3 of water per second (just 10 %[7] of the long-term median water flow) in the main riverbed. This amount is approximately only one third of the lowest ever measured flow (600 m^3/s) in this section of the river. Some decades earlier when a Swiss company was planning hydropower stations for the Mosoni-Danube, a southern side-arm in this area, the Hungarian water authorities strictly stated that this minimal amount should be the overall minimum flow for the main riverbed. Presumably this limit was the main reason that the project was abandoned. The projectors later drastically decreased the desired limit, trying to make their project more attractive. In 1952, a statement by water management experts declared that the minimum water flow in the river bed should be calculated from the lowest permissible groundwater table level, determined by the requirements of agriculture[8] (Bogárdi et al. 1952). However, not only was a scientific investigation into this matter neglected for a long time but the volume of water left in the original river bed was further decreased arbitrarily by half *two times* in the following decades (Vargha 1981). As a result, if the project were now being operated according to the treaty signed by Czechoslovakia and Hungary in 1978, only 50 m^3 of water per second would flow at the bottom of the river bed (except on a few days per year when the river flow exceeds the maximum amount of water that may pass through the turbines).

[6] This information comes from top secret records about a meeting held on 18 April 1953 to discuss the Danube Power Plant in the office of the then deputy Prime Minister Ernö Gerö (published in 1985 in Beszélő, a samizdat journal).

[7] It must be suspected that this simply derives from the use of the decimal notation in continental Europe.

[8] Not to speak about the amount and dynamics of the water flow required by wetland ecosystems of a wide floodplain, interwoven with numerous smaller and larger river branches.

It should be emphasized that the steps that were undertaken to increase energy output by a few percent were made without investigations by the projectors into the increasing risk of their impact on agriculture and wetlands.

The last major redesign involved the addition of a lowland reservoir downstream from Bratislava and the doubling of the power generating capacity of the Gabčíkovo power plant to make the system capable of hydropeaking.[9] This change has made the weaker (least economical) Nagymaros dam indispensable for maintaining the water level required for navigation out of peak time when the whole river is stopped by the upper dam at Gabčíkovo. This redesign did not increase the amount of energy generated (in fact, it decreased it by few percent per year) but it made it possible collect the river water in the reservoir on a daily basis and to operate the turbines 5–6 h per day during peaks in electricity demand. At the time of the redesign the price of electrical energy produced during peak time in the Soviet Union and satellite countries was set to double that produced at non-peak time. This observation showed that the project was a bit more feasible in spite of the additional (and costly) need for turbines and generators with double the capacity and made the projectors more optimistic about the success of their project. However, Emil Mosonyi, the chief engineer of the Nagymaros dam, did not bother investigating issues such as the possible environmental impact of the shallow lowland reservoir, the daily stopping and flushing of the water from the entire river, the clearance of 60 km^2 of floodplain forests, and so on. The practical results of the projectors' self-interests went far beyond the limits suggested by engineering experience and guidelines, and indeed, beyond common sense. No consideration was given to the interests of other stakeholders, like locals who would have to endure the impacts of the transformation of nature.

Underestimation of project costs. In relation to this major redesign, Emil Mosonyi declared that opinions about hydropower utilization should be fundamentally changed in Hungary, and that "our river projects should be regarded as having considerable capacity, the ability to produce a remarkably large amount of peak energy and as absolutely economical, even by international comparison". It may be emphasized—wrote Mosonyi –, "that the most notable technical simplifications and largest cost reductions were achieved during the planning stages of so-called additional investments which were made to protect areas from the water dammed up by the barrages" (Mosonyi 1961). Engineers suggested that irrigation was necessary to balance the negative impacts of the drop in groundwater but the necessary investments costs were never included in the project budget. Such financial tricks, together with the hiding of a considerable portion of the power plant's construction price in the budget of other sectors, created a more favorable impression about the project and contributed to the political decision in 1978 to undertake the Gabčíkovo-Nagymaros project.

Flyvbjerg and his colleagues analyzed 258 public works projects from different geographic regions and historical periods and concluded that the widespread, statistically significant practice of cost underestimation that occurs "cannot be explained by error and is best explained by strategic misrepresentation, i.e., lying" (Flyvbjerg et al. 2002).

---

[9] This involves how the power plant meets changes in demand for electricity (river water is used for energy generation when demand is high and is collected in a reservoir during non-peak periods). This method of operating may have significant impacts on the river ecosystem.

False promises. In spite of all the trickery the Danube megaproject was still too expensive for Hungary. The ultimate financial trick to convince the Hungarian government to give the go-ahead was a promise by the Soviet Union that it would cover a part of the project costs in the form of a loan. However, soon after the Czechoslovak-Hungarian bilateral treaty for constructing the Gabčíkovo-Nagymaros projects was signed, Moscow withdrew this promise.[10] The Hungarian government, feeling cheated, temporarily suspended the construction works on her side in 1980, referring to a shortage of money.

Censoring/editing the negative opinions of experts. Hungarian water engineers made two tours in 1959 and 1962 to study the Grand Canal of Alsace, the political and technical prototype of the Danube canal. The first delegation reported that the construction of the originally planned 120 km long canal had been abandoned halfway through because of the fall in the groundwater table on the German side of the Rhine. This unfavourable report was soon forgotten and a second delegation—of which Emil Mosonyi was a participant—concluded that "according to the experience gained at the already constructed power plants [in the Rhine canal], the insulation of the [Danube] canal, the protection of the surrounding area and the balance of the groundwater can be assured" (Vargha 1981).

In 1982–1983, following the temporary suspension of construction at the Gabčíkovo-Nagymaros project in Hungary, the National Technical Development Committee of the government re-examined the project. This investigation was proposed by the Hungarian Academy of Sciences (HAS) but the projectors controlled the editorial content of the final report. András Lévai, an energy expert and member of HAS, wrote the following about the editors' methods: "[even my] remarks, made always with good intent, were completely ignored or considered just tangentially (some people say), or simply refused. I don't see too much sense raising the questions again because the editorial board with their attitude of 'everything is good just as it is written in the plans' would not consider them anyway" (Lévai 1983). In fact, the projectors did not refrain from more seriously manipulating experts' opinions or interfering with their interests. In a draft report for a meeting of the Hungarian section of the Joint Czechoslovak-Hungarian Regional Planning Committee, Hungarian regional planners wrote that:

> The water agency prepared several alternative plans for decreasing the scale and area of impact caused by the change in the groundwater table. These are more favorable [...] than the original plan, but they are not suitable for maintaining the present level of agricultural production nor for protecting the floodplain forests.

The text was *edited* by the water engineers, with some parts deleted and others added. The result was a final text with a meaning completely different to the original[11]:

> The water agency prepared several alternative plans for decreasing the scale and area of impact caused by the change in the groundwater table. Among them some solutions came to the fore that will serve to maintain the present level of agricultural production and protect the floodplain forests.

---

[10] Further research may clarify the details of this story.

[11] A facsimile of the typed text with hand-written 'editions' was published in a book about the decision of the International Court of Justice on Gabčíkovo-Nagymaros (Vargha 1997).

Fait accompli. The projectors committed everything to make their projects attractive; promising progress, development, and public welfare to gain political support. They made many attempts to silence their opponents but were never able to feel that they were standing on firm ground as they were fully aware of the truth. Instead, they tried to convince decision-makers and the public that even if there were some (of course, minor) difficulties, it was too late to reconsider the decision; it would now be better to complete the project than not.

In the Gabčíkovo-Nagymaros case, the policy of promoting a *fait accompli* started in 1953 with the above-described attempt to gain the Hungarian government's immediate agreement to divert the Danube into the left-side canal that would be built on Czechoslovak territory.

A few years later (in 1958), the projectors achieved partial success. The Hungarian Prime Minister, Ferenc Münnich, signed an agreement in Prague about the joint construction of a dam and hydroelectric power plant at Visegrád—and agreed to continue combined planning for the utilization of the upper section of the Danube. However, the projectors' hopes were soon extinguished for two main reasons: Czechoslovakia was reluctant to take part in the construction without Hungary's consent to divert the upper section of the river to her territory, and Hungary later found the Visegrád power plant unattractive when compared to other power generation options.

In 1980, when Hungary temporarily suspended construction for financial reasons, Czechoslovakia tried to force Hungary to continue the project. To create an impression of *fait accompli* the uppermost fertile layer of soil (together with the growing wheat) alongside the planned side-canal was removed and a small part of the Gabčíkovo dam was quickly built (a concrete wall separating the entrance of the navigation locks from the water flow to the turbines). The destruction of the land and the construction of the 16 m high wall were part of an attempt to convince uninitiated politicians about the irreversibility of the project.

Fait accompli is a frequently-used argument by projectors and sometimes even project plans are referred to this way. In the case of the Canadian James Bay project the Quebec provincial Government did not consult with local people when the plans were drawn up and they "presented the project as a fait accompli, telling the Indians that the scheme would go ahead regardless of their opposition" (Goldsmith and Hildyard 1986). In debates with environmental groups projectors and their allied politicians and authorities frequently proclaim that the project in question should be regarded as a fait accompli and any investigations and environmental impact assessments will only result in minor modifications (Fearnside 1989; Vagholikar 2011).

An extreme example of the *fait accompli* approach is the closing of the Danube River downstream from Bratislava in 1992; the projectors response to the efforts of environmental groups to stop the construction works (Fürst 2003; Galambos 1993). This attempt was a gamble with the safety of the surrounding inhabited area as the whole river was closed with locks only half-ready to manage flood and ice management demands (OVIBER Kft. 1994). In order to placate the response to the closure, 1 week later—in a minute signed in London by delegates from Czechoslovakia, Hungary and the European Commission—the projectors "taking into account the risk of damage to existing structures including navigation, of ecological damage to the region and of flooding" guaranteed "to maintain the whole traditional quantity of water into the whole old riverbed"[12] and "refrain from operating the [Gabčíkovo] power plant" (I.L.M. 1993). However, the engineers did not comply with this agreement for one minute, undoubtedly having the silent consent of supportive politicians.

3 Dams or Rivers?

Paul R. Josephson, in his book about brute force technology and the transformation of the natural world, writes that "the most troubling aspect of brute force technologies is that they require brute force politics for full effect. This is because financial institutions, engineering firms, state bureaucracies, and specialists themselves develop a strong interest in seeing their designs fully unfurled. There are no half measures with brutal force technologies" (Josephson 2002). This observation characterizes the damming of rivers, analyzed by Josephson in the first chapter of the book, entitled "Pyramids of Concrete: Rivers, Dams, and the Ideological Roots of Brute Force Technology".

However, in most cases concerning hydropower "half measures" are neither reasonable nor sustainable solutions in any case. With the Gabčíkovo-Nagymaros project the judgment of the International Court of Justice involved a typical "half measure". The Court accepted the status quo (i.e. the *fait accompli*) involving the ongoing operation of the Gabčíkovo power plant in the side canal, but it also obliged the parties involved to ensure that nature was protected. The judges envisioned reconciling economic development with environmental protection, stating that "the Parties together should look afresh at the effects on the environment of the operation of the Gabčíkovo power plant. In particular they must find a satisfactory solution for the volume of water to be released into the old bed of the Danube and into the side-arms on both sides of the river" (International Court of Justice 1998). The planning process culminated in a final proposal to leave only 2.5 % of the original water in the Danube and to divert 97.5 % of it to the turbines. The failure of Slovak-Hungarian negotiations "to find a satisfactory solution" in the 16 years following the Court's judgment (and mitigate the environmental impacts in the area surrounding the river-bed during this period) shows that the Gabčíkovo-Nagymaros project is not an exception to the rule. The problem remains that even sharing the water roughly fifty-fifty between the turbines and the original riverbed would render the operation of the power plant completely uneconomical but would not protect nature. However, the current situation is not balanced at all; according to a 'provisional' agreement the power plant operates using about 80 % of the water from the Danube. The projectors are the winners and the ecosystem is the loser.

[12] Footnote in the original text: "whole" means not less than 95 % (I.L.M. 1993).

Philippe Sands, professor of international law at University College London, criticized the Court, saying that “in cases such as this, where environmental concerns lie at the heart of the dispute, the Court has, thus far to date, declined to avail itself of the power set out in Article 50 of the ICJ Statute and retain its own scientific experts to help unravel the complexities that are inherent in arguments of a scientific and technical nature” (Sands 2010). The Court concluded that “it is not necessary in order to respond to the questions put to it in the Special Agreement[13] for it to determine which of those points of view is scientifically better founded” (International Court of Justice 1998). The remarkable ignorance that ensued from this decision resulted in the Court prescribing an impasse to the parties concerned. But in reality the deadlock is not balanced at all; as regards the reservoir, the side-canal and the Gabčíkovo power plant the projectors have indeed achieved their *fait accompli*.

Josephson explains such situations by referring to the great institutional momentum acquired by brute force technologies and states that “there is no going back or cutting losses, and rarely do we admit failure and disassemble them—for example, by breaching a dam or saying no to a project in process” (Josephson 2002). However, a (partial) counterexample is the Gabčíkovo-Nagymaros project itself; environmental groups managed to stop the construction of the second barrage in 1989 (which had already started at Nagymaros where the site was surrounded by a coffer dam and the river already diverted into a provisional bed). A few years later the coffer-dam was removed and the river was restored to its original bed. In this section the Danube River now flows freely again. Here, the Danube Circle lived up to its goal, encapsulated in the motto of the group—borrowed from Comenius[14]: “*Omnia sponte fluent, absit violentia rebus*” (Let everything flow free, violence shall be afar from things).

References

Bogárdi, J., et al. 1952. *Vélemény a Bratislava-Visegrád közötti Duna-szakaszon tervezett vízlépcsők hatására bekövetkező hordalék- és jégviszonyokról* (Opinion of the impact of dams planned for the Danube’s Bratislava-Visegrád section on sediment and ice conditions, Manuscript).

Bogdánfy, Ö. 1914. *A vízerő: mérnöki kézikönyv* (Hydropower: Engineering handbook). Budapest: Budapest Magyar Mérnök- és Építész Egylet.

Carkoglu, A., and M. Eder. 2001. Domestic concerns and the water conflict over the Euphrates-Tigris river basin. *Middle Eastern Studies* 37 (1): 41–71.

Chioc, M. 2002: *The Rhine: An eco-biography, 1815–2000*. Seatle: University of Washington Press.

Descargues-Grant, M. 2011. Swift and the Ruin(s) of history. *Études Épistémè* 19 (1): 113–128.

[13] This agreement relates to questions submitted to the Court for judgment by Hungary and Slovakia.

[14] Comenius was a teacher, educator and writer in the XVI. Century. See: http://en.wikipedia.org/wiki/John_Amos_Comenius.

Fearnside, P. M. 1989 "Brazil's Balbina Dam: Environment versus the legacy of the pharaohs in Amazonia. *Environmental Management* 13 (4): 401–423.

Fitzmaurice, J. 1996. *Damming the Danube: Gabcikovo/Nagymaros and post-communist politics in Europe*. Boulder: Westview Press.

Flyvbjerg, B., M. S. Holm, and S. Buhl. 2002. Underestimating costs in public works projects: Error or lie? *Journal of the American Planning Association* 68 (3): 279–295.

Fürst, H. 2003. The Hungarian-Slovakian conflict over the Gabcikovo-Nagymaros Dams: An analysis. *Intermarium* 6 (2): 1–15.

Galambos, J. 1993. An international environmental conflict on the Danube: The Gabcikovo-Nagymaros dams. In *Environment and democratic transition,* ed. A. Vari and P. Tamas, 176–226. Dordrecht: Springer.

Goldsmith, E., and N. Hildyard, eds. 1984. *Social and environmental effects of large Dams*. Volume 1. San Francisco: Sierra Club Books.

Goldsmith, E., and N. Hildyard, eds. 1986. *The social and environmental effects of large Dams. Volume 2: Case studies*. San Francisco: Sierra Club Books.

Görbedi, M. 1989. *1020 nap az őrtornyok árnyékában: A tiszalöki fogolytábor története* (1020 Days in the Shadow of Watch-towers: The History of the Tiszalök Prisoner Camp). Tiszalök: Költségvetési Üzem.

Gregory, P. R., and V. Lazarev. 2003. *The economics of forced labor: The Soviet Gulag*. Stanford: Hoover Institution Press.

Gruner, W. 2006. *Jewish forced labor under the Nazis: Economic needs and racial aims, 1938–1944*. Cambridge: Cambridge University Press.

Jolánkai, Gy. 1957. A Tisza-völgy északi része vízhiányának pótlása a Dunából átvezetett víz dombvidéki tározásával. *Vízügyi Közlöny* 3 (1): 3–21.

Josephson, P. 2002. *Industrialized nature: Brute force technology and the transformation of the natural world*. Washington, DC: Island Press.

Higgins, I. 1994. *Swift's politics: A study in disaffection*. Cambridge: Cambridge University Press.

I.L.M. 1993. Czech and Slovak federation-european commission-hungary: London agreement on the Gabčíkovo-Nagymaros Project. *International Legal Materials* 32 (5): 1291–1292.

International Court of Justice. 1998. Judgment in case concerning the Gabčíkovo-Nagymaros Project. *International Legal Materials* 37 (1): 162–242.

Kiss, G. 1947. TVA on the Danube? *Geographical Review* 37 (2): 274–302.

Lévai, A. 1983. Opponensi vélemény (Examiner's report). In *A Duna távlati komplex hasznosítása* (The long-term complex utilization of the Danube). Budapest: MTA-OMFB.

Lunstrum, E. 2011. An uncomfortable fit? Transfrontier parks as MegaProjects In *Engineering Earth: The Impacts of Megaengineering Projects,* ed. S. D. Brunn, 1223–1242. Dordrecht: Springer.

McCully, P. 2001. *Silenced rivers—The ecology and politics of Large Dams*. London: Zed Books.

Mosonyi, E. 1948. Magyarország vízerői (Hydropower of Hungary) *Vízügyi Közlöny* 30 (Separatum): 1–15.

Mosonyi, E. 1951. Vízerő-gazdálkodásunk időszerű feladatai (Current Tasks of our hydropower management). *Magyar Tudományos Akadémia Műszaki Tudományok Osztályának Közleményei,* 515–540. Budapest: MTA.

Mosonyi, E. 1961, March. Új elgondolások vízerőink kihasználásában (New concepts about the utilization of our hydropowers). *Hidrológiai Tájékoztató* 17–18.

OVIBER Kft. 1994. *A szlovákiai C-variáns vízépítő mérnöki értékelésének végleges szakvéleménye* (The final expert opinion on the hydraulic engineering evaluation of the Slovakian Variant C) Budapest: OVIBER.

Parasuraman, S., and S. Sengupta. 2001. World commission on Dams: Democratic means for sustainable ends. *Economic and Political Weekly* 36 (21): 1881–1891. http://www.jstor.org/stable/4410669.

Rainer, J. 2002. *The new course in Hungary in 1953*. Cold War International History Project, Working Papers No. 38, Washington DC: Woodrow Wilson International Center for Scholars.

Ratcliff, J. 2012. Art to cheat the common-weale: Inventors, projectors, and patentees in English satire, ca.1630–1670. *Technology and Culture* 53 (2): 337–365.

Record of Meeting. 1985. Befejezetlen múlt: Gerő Ernő a vízlépcsőről (Unfinished Past: Ernő Gerő about the Dam). *Beszélő* 1 (14): 20–28.

Reilly, P. 1982. *Jonathan swift: The brave desponder*. Carbondale: Southern Illinois University Press.

Rogers, P. 1975. Gulliver and the engineers. *The Modern Language Review* 70 (2): 260–270. http://www.jstor.org/stable/3724280.

Rothschild, J., and N.M. Wingfield. 2000. *Return to diversity: A political history of east central Europe since World War II*, 3rd ed. New York: Oxford University Press.

Sajó, E. 1931. *Emlékirat vizeink fokozottabb kihasználása és újabb vízügyi politikánk megállapítása tárgyában*. (Memorandum about greater utilization of our waters and establishing our new hydropolitics. Manuscript)

Sands, P. 2010. Water and international law: Science and evidence in international litigation. *Environmental Law & Management* 22 (4): 151–161.

Swift, J. 1726. *Gulliver's travels. Webster's Thesaurus Edition (2005)*, Las Vegas: ICON Group International, Inc.

Vagholikar, N. 2011. Dams and Environmental governance in North-East India. In *India infrastructure report 2011: Water: Policy and Performance for Sustainable Development,* ed P. Tiwari and A. Pandey 360–369. Oxford: Oxford University Press. http://indiasanitationportal.org/sites/default/files/IIR-2011_1.pdf#page=401.

Vargha, J. 1981. Egyre távolabb a jótól (More and more distant from that which is good). *Valóság* 24 (11): 1–23.

Vargha, J. 1984. (Pseud.: Kien Péter) "A Nagy Szlovák Csatorna" (The Great Slovak Canal). *Beszélő* 9 (1): 1–19.

Vargha, J. 1997. Vízerő és politika (Hydropower and politics). In *A hágai döntés* (The decision at The Hague), ed. Janos Vargha, 221–287. Budapest: Enciklopédia Kiadó.

Weijnen, T. 2010. *Governance of large infrastructures: The cases of the canals of King Willem I, the Suez Canal and the Rhine-Main-Danube waterway. An application of new institutional economics*. Enschede: University of Twente. http://purl.org/utwente/doi/10.3990/1.9789036530729. Accessed 30 July 2013.

Williams, R.C. 1995. The environment, borders, and politics: The significance of central Europe's Gabcikovo-Nagymaros hydroelectric project before and after 1989. *Middle States Geographer* 28 (2): 242–255.

Zinke, A. 2005. The hydropower station Gabčíkovo: Deficits in hydrology (Sediment Transport, Groundwater) and biology. In *Hydrology and limnology–another boundary in the Danube River Basin. Report on the International Workshop organized by the International Association for Danube Research (IAD) and sponsored by the UNESCO Venice Office,* 49–59. Petronell, Austria, 14–16 October 2005.

Janos Vargha is an environmentalist and photographer. A graduate from the József Attila University, Szeged, with Master's degree in biology. From 1981 he writes articles and delivers lectures about the environmental issues of the water management and constructions. He is a founder of the Danube Circle, the first independent environmental protest group in the former communist block, recipient of the Right Livelihood Award, the alternative Nobel-prize in 1985. In 1990 he was awarded with the Goldman Environmental Prize. He is editor of a book, published in 1997, about the decision of the International Court of Justice in the Gabčíkovo-Nagymaros hydropower project's case, and author of its chapter about hydropolitics. From 1998 he was the chief environmental advisor of the Hungarian government. In 2000 he resigned from this position. Now he works as a photographer and web developer

Entrepreneurship Inspired by Nature

Nel Hofstra

In modern societies Nature has been overexploited by firms. This is partly because modern societies favor capital and labor at the expense of Nature and because Nature is considered to be a commodity. Capital and labor have become the dominant pillars of business strategies that lead to the exploitation and destruction of Earth's systems.

Our perception of Nature determines how we assess environmental problems, how we assess and review the status of critical ecosystems and how we try to solve these problems. Sustainable entrepreneurs create innovative products and services that can reconnect humans with Nature. It is not how to exhaust and exploit natural resources but how to renew natural services and functions that is the biggest challenge for innovation. This paper is written to give an answer to the question: *How can entrepreneurs foster the development of regenerative eco-innovations?*

1 Models in Economics

Steve Keen debunked the myth that there is no alternative to dominant neoclassical economic theory. He also declared that none of the main alternative theories (Austrian, Post-Keynesian, Evolutionary or Complexity theory) are strong enough to serve as appropriate economic theories for the twenty-first century (Keen 2011). All of the alternative theories are based on neoclassical concepts and do not go beyond the standard economics curriculum.

The ideology of "the invisible hand" played a key factor in the political and social transformations of the late eighteenth and early nineteenth centuries (Keen 2011, p. 172). The financial crisis accelerated the number of complaints against the dominant economic system.

N. Hofstra (✉)
Erasmus University Rotterdam, Rotterdam, The Netherlands
e-mail: nelhofstra@gmail.com

L. Zsolnai (ed.), *The Spiritual Dimension of Business Ethics and Sustainability Management*, DOI 10.1007/978-3-319-11677-8_13

The volume of the call for change in economic theory seems to be in opposition to the rather weak response that the ecological crisis has aroused. Financial threats put pressure on people directly. Ecological threats do so much less, as we have lost our direct connection with Nature. Capital is considered to be more important than Nature. Humans are also considered to be more significant than the biosphere (plants, trees, animals and all living organisms). We may be trying to solve the financial crisis with financial means but we are not trying to deal with the ecological crisis using natural solutions.

2 Environmental Worldviews

Ideas about the relationship between man and Nature can be divided into four categories (Proops 1989):

- 'Undisturbed' Nature
- Man as part of Nature
- Man as creator of Nature
- Creative, self-sustaining Nature (Gaia)

A comparison of environmental worldviews sheds light on the different value orientations employed in economics and management (Table 1).

To make the required shift from an anthropocentric to a more ecocentric approach requires an examination into the consequences of such a change on entrepreneurship.

Table 1 Comparisons of environmental worldviews. (Miller 2007)

Planetary management	Stewardship	Environmental wisdom
We are separate from the rest of Nature and can manage Nature to meet our increasing needs and wants	We have an ethical responsibility to be caring managers or stewards of the earth	We are part of and totally dependent on Nature and Nature exists for all species
Because of our ingenuity and technology we will not run out of resources	We will probably not run out of resources, but in any case they should not be wasted	Resources are limited, should not be wasted, and are not all for us
The potential for economic growth is essentially unlimited	We should encourage environmentally beneficial forms of economic growth and discourage environmentally harmful forms	We should encourage Earth-sustaining forms of economic growth and discourage Earth-degrading forms
Our success depends on how well we manage Earth's life-support systems, mostly for our benefit	Our success depends on how well we manage Earth's life-support systems for our benefit and for the rest of Nature	Our success depends on learning how Nature sustains itself and integrating such lessons from Nature into the ways we think and act

3 Sustainable Entrepreneurship

The sustainable entrepreneur plays a key role in the dynamics of the economic market system and the revitalization of eco-systems. Through profiting from market disequilibrium sustainable entrepreneurs endeavor to innovate in their responses to the customer's wishes for a better world. "The ability to look ahead and anticipate the ways business can harness new innovation is becoming a fundamental skill for businesses that want to accelerate through this shift and come out ahead. Smart businesses that regard change as an opportunity will not only adapt, they'll thrive" (Steffen 2010). Entrepreneurs in the future will be confronted with the need to make the transition from a focus on profit to a focus on value, and from competition to cooperation. The limits of the carrying capacity of the Earth may restrain further economic growth and require new solutions to problems. Old economic and business models do not meet the criteria of being eco-efficient and eco-effective.

Eco-efficiency means generating economic value through successfully addressing ecological issues and regulations and meeting human needs while being as resource-efficient as possible.

Eco-effective solutions involve the delivery of goods and services which meet human and natural demands and improve quality of life. Such goods and services are ecologically appropriate and may be inspired by Nature. They have the potential to meet still unknown, future societal and ecological problems (Hofstra and Huisingh 2014, p. 8).

Sustainable entrepreneurs are faced with the need to be cost-efficient and risk-seeking on the one hand and flexible and innovative on the other. They may have to reframe their strategies thoroughly to become more ecologically effective. Innovations that help society to live in a more ecologically-effective manner can multiply when they are used as a strategy for facing the challenges that Nature offers.

4 Eco-Innovation as Related to the Human-Nature Relationship

Eco-innovation refers to the use of ecological knowledge to bring forth ecological progress (Fussler and James 1996). Many firms have been developing ecological innovations; mostly products and processes that reduce environment-related costs. However, eco-efficiency does not guarantee environmental effectiveness. The concept of eco-innovation is thus evolving. In this paper we focus on the concept of Nature and the steady shift from an anthropocentric to a more eco-centric approach in sustainable entrepreneurship and innovation which is needed to preserve the vitality of the whole system.

Nature as a concept is often taken for granted as a common body of knowledge. Our perspective on Nature determines what we see as Nature or 'the environment'. Different metaphors (like "Mother Earth", "Spaceship Earth", "Production

Resources" or "Natural Capital") reflect some of these perspectives. Categorization of the relationships between humans and Nature (based on an in-depth analysis of historical and philosophical developments of ideas pertaining to human and Nature interactions) has led to the construction of a typology of eco-innovations (Hofstra and Huisingh 2014) In this paper we focus on the regenerative dimension of eco-innovation and the meaning of this for sustainable entrepreneurship.

5 Regenerative Development

Regenerative development affirms and supports living systems (human and biotic). It represents a way of moving towards sustainability that has deep meaning and significance (Ministry 2009). Invention and design with a regenerative character create new opportunities for the practice of eco-innovation. Such products and services recognize the interconnectedness and unity of all life, physical as well as spiritual. They acknowledge Nature as both mentor and teacher because understanding the mutual relationship between man and Nature is essential for the survival of the whole.

Regenerative innovations 'submit' to systems that restore, renew or revitalize their own sources of energy and matter. This process of regeneration should stimulate the Earth's repair of itself and the renewal of its functioning. Our knowledge of Nature is still in its infancy, or is entirely lacking. The concept of dualism between man and Nature is ubiquitous to Western thought and has made it possible for humans to consider Nature as an objective reality. This way, nature can be studied as an object, or a commodity; a mode of thinking that has been incorporated into current economic models. According to Prigogine, the endeavor of modern man to understand Nature has been based on the idea of control (Prigogine 1996).

To bring forward a new relationship with Nature we need new economic structures to promote tools which are appropriate for use 'in' Nature. Ecuador is the first country to have changed its constitution to officially recognize that organisms in the eco-system (Nature) have "the right to the maintenance and regeneration of their vital cycles, structures, functions and evolutionary processes". Being committed to the universe inspires people to create new designs. The Ministry for the Environment of New Zealand focuses on a regenerative development approach. The departure away from promoting eco-efficiency is a significant first step but is not effective enough for the long term restoration scenarios to be realized.

Regenerative developments require fundamentally rethinking our connection with Nature. A low level of connectivity, separation and disassociation between the human race and the universe hampers a holistic approach to our environmental problems. The required transformation from a kind of development that degenerates ecosystems to one that regenerates their capacity to flourish is illustrated in Table 2.

A whole-systems or holistic approach is a virtuous cycle where positive action is reinforced through feedback mechanisms. Values are expressed qualitatively rather than quantitatively which requires alternative systems for measuring economic value. An example of this approach is the Gross National Happiness (GNH) index

Table 2 Business-as-usual compared to regenerative development. (Ministry 2009, p. 8)

Conventional approach	Eco-efficiency	Regenerative development
Little or no consideration is given to the environmental impact of the design	Green design: Does not challenge current production methods or consumption patterns that have negative environmental impact	Sees humans, human developments, social structures and cultural concerns as an inherent part of ecosystems
Designs generally aim to meet minimum legal requirements for the lowest first-cost price	Minimizes energy use, pollution and waste (termed 'less bad' design)	Questions how humans can participate in ecosystems using development to create optimum health
A rapidly expanding segment of business-as-usual is 'green' and moving towards 'sustainable'	Sustainable design: Achieves neutral environmental impact and maximum efficiency	Seeks to create or restore capacity of ecosystems and bio-geological cycles to function without human management
		Understands the diversity and uniqueness of each place (socially, culturally and environmentally) as crucial to design
		Sees the design process as ongoing, indefinite and participatory

in Bhutan which has its origin in Buddhist principles. This set of indicators was developed in an effort to identify and evaluate quality of life in more holistic terms than simply using Gross Domestic Product (GDP).

6 Applying Regenerative Strategies

The destruction of Nature for human gain is a degenerative pattern which must be transformed into a regenerative strategy. Palliative approaches in environmental regulation have not satisfactorily dealt with the essential structure of one-way-flows that we have not only built into our apparatus of production but into the very design of fundamental sources of sustenance (Lyle 1994). Lyle offers a preliminary list of regenerative design strategies, which is anything but definitive (Lyle 1994, pp. 38–48) (See Table 3).

Continuous self-renewal is the key principle to designing regenerative systems. Regenerative solutions are in their infancy. It is Nature itself that delivers the knowledge of understanding. There is not yet a coherent body of knowledge about this approach. The *Center for Regenerative Studies* in Pomona, California (related to the Polytechnic University) is the first community to create a regenerative system in which education is the motive force and the ecosystem is seen as a medium for understanding and figuring out the natural and social world. Students are provided with an interdisciplinary master's program about meeting the environmental challenges of the twenty-first century through regenerative approaches and are encourage to

Table 3 Applying regenerative strategies

1. Letting Nature do the work (passive solar heating and cooling systems, use plants for micro-climate control, use thermal storage properties and induce natural air movement within buildings, employ landforms to guide the flow of water, use predator species and companion planting as primary means of pest control)
2. Consider Nature to be both model and context (follow natural patterns of resource distribution, emphasize water efficiency; reuse and recycling)
3. Aggregate, don't isolate (functions like energy generation, shelter, water management, food production and waste recycling)
4. Seek to manage for optimum levels and multiple functions, not maximum or minimum levels (operate interactively and in equilibrium, promote mutually-beneficial multi-functionality)
5. Match the technology to the need (for example, concentrating solar collectors for solar food-drying racks)
6. Use information to replace power (observe Nature's operations, monitor quality)
7. Provide multiple pathways (no monopoly for service systems, food production, public utilities)
8. Seek common solutions to disparate problems (for example, roofs of buildings can serve for growing food, collecting water, heating water, generating energy, circulation, and activity areas as well as for protection from the weather)
9. Manage storage as a key to sustainability (for example, thermal-mass materials provide heat storage in buildings and for food storage, drying, smoking and canning)
10. Shape form to guide flow (building forms intercept maximum levels of solar radiation, with means for blocking it out when necessary)
11. Shape form to manifest process (technological devices are in visible locations wherever possible and hill top wind and solar generators are major features in the landscape)
12. Prioritize for sustainability (for example systems that provide means for on-site recycling)

think about how to best create renewable systems that support communities, using waste as a resource and Nature as a model.

7 Entrepreneurship Inspired by Nature

Business according to Nature implies systems thinking, cyclical thinking and spiritual thinking which is focused on seeking the catalytic interfaces that create opportunities for the revitalization of our social, economic and eco-systems. Entrepreneurs who are inspired by Nature can become capable of learning to work with natural uncertainty. Sustainable entrepreneurs inspired by Nature can become true innovators in building economies and markets. Nature then will become a stakeholder of the most importance. The recognition of opportunities involves the identification of Nature's patterns through observation, discovery, admiration and exploration.

There are already some good examples of products and services which have been designed with these principles in mind. Regenerative medicines that are used for replacing or regenerating human cells or organs to their customary functioning are starting to replace conventional medicines used by the pharmaceutical industry. Bone regeneration is fostered by mimicking natural conditions that favor bone growth. Solar cells are inspired by leaves and the pot-in-pot refrigerator by thermodynamics (Steffen 2010). Obsolete or as yet undiscovered knowledge about food processing will transform the food industry into a slow food movement. Fire retardant boards made of water, flour, oyster mushroom spores and prelate. A cardboard bottle that will decompose within six weeks after use. Displays powered completely by photovoltaic cells. A cargo vessel with a supplementary propulsion system. Floors made from cow manure. The use of fog, rain and dew as sustainable water resources in arid regions. Plastic from plants. A dishwasher that cleans and sterilizes using steam, instead of chemicals. Regenerative counter-top water conditioners, hydro cyclones, incinerators and batteries. Radios that don't need batteries or any other external electricity supply, just solar cells or a winding handle. Lexus hybrid cars use a technology called regenerative braking where kinetic energy is converted into electric energy. Mobile phones that are recharged using human warmth and fitness schools that recover the energy burnt off by their customers are illustrations of how to regenerate energy.

To excel as an entrepreneur, working hand in hand with Nature, five distinctive principles must be followed (Hofstra 2007):

- Anticipate and transform unforeseen threats caused by natural resources into opportunities, where Nature is considered as a teacher and mentor
- Meet the multiple interests of business, society and Nature
- Search for long term economic efficiency combined with ethical concerns and ecological effectiveness
- Explore, don't exploit Nature
- Create business value—in the broader sense of economic, ecological, moral, social and spiritual value

The cradle-to-cradle concept promotes the use of waste as a new (re)source for manufacturing in the bio and technosphere (Braungart 2006). 'Bioneers' is a movement that provides novel ways to join forces in confronting and mimicking design and innovation in a life-conducive way (Beynus 2002). Regenerative developments are affirmative contributions to living systems (human and biotic) and present entrepreneurial opportunities. They require constant adaptation and the embedding of growth with a deeper meaning and significance for all involved. This way of thinking will opens the way to a completely novel realm for entrepreneurs and will simultaneously awaken us to the magnitude of opportunities, instruction and wonder that our Earth can provide.

References

Beynus, J. M. 2002. *Biomimicry: Innovation inspired by nature*. New York: Harper Collings Publishers Inc.

Braungart, M. 2006. Cradle-to-cradle design: Creating healthy emissions; a strategy for eco-effective product and system design. *Journal of Cleaner Production* 15 (1): 1337–1348.

Fussler, C., and P. James. 1996. *Driving eco-innovations: A breakthrough discipline for innovation and sustainability*. London: Pitman Publishing.

Hofstra, N. 2007. Sustainable entrepreneurship in dialogue. *Progress in Industrial Ecology—An International Journal* 4 (6): 495–514.
Hofstra, N., and D. Huisingh. 2014. Eco-innovations characterized: A taxonomic categorization for assessing the relationships between humans and nature. *Journal of Cleaner Production* 66 (5): 459–468.
Keen, S. 2011. *Debunking economics; the naked emperor dethroned*. London: Zed Books.
Lyle, J. T. 1994. *Regenerative design for sustainable development*. Pomona: California State Polytechnic University.
Miller, G. 2007. *Living in the environment: principles, connections and solutions*. Belmont: Thomson Brooks/Cole.
Ministry. 2009, *Towards a sustainable future: Adopting a regenerative approach to development*. Wellington: The Government of New Zealand.
Prigogine, I. 1996. *The end of certainty; time, chaos and the new laws of nature*. New York: The Free Press.
Proops, J. 1989. Ecological problems, rationale and problem areas. *Ecological Economics* 1 (1): 59–76.
Steffen, A. 2010. *World changing. A user's guide for the 21st century*. New York: Sagmeister Inc/Abrams.

Nel Hofstra was born in 1954 in Leeuwarden, the Netherlands. She has a doctorate in economics and marketing and business sociology from the Erasmus University Rotterdam. She is lecturer in the Department of Business Economics, Faculty of Economic Sciences at Erasmus University Rotterdam. She teaches Entrepreneurial Marketing, Sustainable Entrepreneurship and Business & Society. She is a member of the Business Ethics Faculty Group of CEMS and doing research in eco-innovations and entrepreneurship, focusing on the concept of the role of Nature. She wrote several publications in this field.

Part IV
Spiritual-Based Leadership in Business

Spiritual-based Leadership: A Paradox of Pragmatism

Peter Pruzan

Although far from mainstream, the concept of spiritual-based leadership is emerging as an inclusive and yet highly personal approach to leadership that integrates a leader's *inner* perspectives on identity, purpose, responsibility and success with her or his decisions and actions in the *outer* world of business—and therefore it is also emerging as a significant framework for understanding, practicing, communicating and teaching the art and profession of leadership.

Spiritual-based leaders are nourished by their spirituality, which is a source within them that informs and guides them. They search for meaning, purpose and fulfillment in the external world of business and in the internal world of consciousness and conscience. Their external actions and their internal reflections harmonize so that rationality and spirituality are mutually supportive (Pruzan 2011).

But these characteristics are considered by many in the academic community as being 'fluffy', and they argue that there is a need to operationalize this perspective on leadership so that it can be subject to rigorous empirical research and evaluated as though it were a traditional academic domain.[1] This leads to the paradox referred to in the title.

While there is an increasing interest in leadership based on a spiritual foundation, the greater the attempts to operationalize this perspective on leadership and make it 'scientific', the more this may lead to its self-eradication. Conversely, the greater the appreciation of its metaphysical basis, the more spiritual-based leadership will thrive and contribute to both the individual leader's development and fulfillment as well as the development and fulfillment of those affected by his or her leadership.

[1] This is not to say that empirical research is not vital for the development of the field. I can refer to my own primarily qualitative research, carried out among top leaders from 15 countries in six continents (Pruzan and Pruzan Mikkelsen 2007, 2010).

P. Pruzan (✉)
Copenhagen Business School, Frederiksberg, Denmark
e-mail: pp.mpp@cbs.dk

Sri Sathya Sai Institute of Higher Learning, Prasanthi Nilayam, Andhra Pradesh, India

L. Zsolnai (ed.), *The Spiritual Dimension of Business Ethics and Sustainability Management*, DOI 10.1007/978-3-319-11677-8_14

Perhaps an analogy can be made to the concept of 'love'—which is also a term frequently referred to by spiritual-based leaders (Pruzan and Pruzan Mikkelsen 2007, 2010). While we are able to feel and give expression to love and the more poetically inclined are able to create verbal imagery that resonates with our inner sense of love, attempts to define, categorize, measure and optimize this essential quality transform it from its essence, a feeling of unity, to a concept.

Lessons can be learned about this paradox from the developments in other, closely related perspectives on leadership. There is a striking similarity in developments in the broad fields of Business ethics, CSR and Sustainability, referred to collectively in the sequel as B-C-S. In the relatively short period of time of roughly 30–40 years, all three fields have been characterized by a movement from philosophical reflection and critical perspectives on organized commercial activity, to disciplines characterized by their own vocabularies, measures of performance, university curricula, professional journals, consultancies, non-governmental organizations, international organizations and reporting systems. Depending on one's own aspirations as to the 'common good', these developments may be encouraging or alarming.

The more optimistic amongst both leadership theoreticians and practitioners, and I am one of them, refer to a significant new orientation amongst both business and political leaders whereby they include B-C-S dimensions in their rational justifications of decisions. In addition, reference can be made to an increasing awareness of the significance of these dimensions by corporate stakeholders and the citizenry.

The more pessimistic, and I belong to that group as well, refer to the fact that the operationalization of the original critical philosophical perspectives has resulted in their being colonized by and reduced to elements in an economic rationality; they no longer are important in their own right but are looked upon as instruments to protect corporate reputation and to promote economic growth. In particular, those who are critical with respect to the functioning of the globalized capitalistic system with its almost alchemistic transformation of *everything*, including intangibles we value (such as justice, beauty, serenity, love and respect) but which do not readily lend themselves to economic measurement, deplore the rapid developments from the metaphysical to the physical, from the intrinsic to the extrinsic. Philosophically deep concepts such as ethics, responsibility and sustainability, each pregnant with significance for reflection on the human condition, have been operationalized and reduced to indicators and variables that can be measured, evaluated and reported.

As can be seen in this volume, a result has been an increasing call for the spiritualization of ethics, responsibility and sustainability. Inherent in this call is the aspiration for a re-enchantment of the fields so they develop in accord with their philosophical and humane bases. This is not a naïve call for a reactionary dismantling of these fields, but for a critical and caring re-vitalization of the perspectives and methods that have characterized their developments. In my own writings (Pruzan 2009; Pruzan and Pruzan Mikkelsen 2007, 2010), I have referred to this call as a search for spiritual-based leadership—as a search for a deeper, spiritually founded well-spring of the rationality and morality that implicitly characterize the fields of B-C-S.

However, and this underlies the paradox referred to here, the evolution of a spiritual-based leadership implies not just a transformation of the teaching and the practice of leadership, but also, and more fundamentally, the transformation of the individual leader's mind-set. What is required is a consciousness that resonates with a conviction that a precondition for the long-term success of purposeful, organized mercantile activity is spiritual-based leadership and not just the pursuit of material gain.

By this I do not imply that it is not important for the future welfare of this globe and its inhabitants to improve research in, the teaching of, and the implementation of tools based on B-C-S as we know these fields today. It will continue to be important for organizations of all kinds and at all levels to develop their own perspectives on ethics, responsibility and sustainability. However, unless leaders in the academic community as well as in corporate, consulting, regulating and advisory bodies actively focus on motivating and inspiring decision makers to supplement their traditional success criteria with spiritual-based perspectives, business ethics will continue to deal more with *un*ethics than with ethics; CSR will continue to emphasize the protection of corporate reputation and success rather than responsibility to broader constituencies; and sustainability will continue to focus on promoting conditions and technologies that enable business growth rather than the maintenance and improvement of the 'common good'—including embracing constituencies that lack voice, such as nature and the yet-unborn.

At a more down-to-earth practical level and in contrast to the analyses that characterize present day managerial and scientific rationality, this will imply that our theoreticians will face the challenge of developing vocabularies, perspectives and research methods that can support leadership that is spiritually-based. Instead of a focus on deliberate and willed action that is considered to be the result of logical generalizations and prescriptive principles, this implies a focus on the emancipation and empowerment of inner guidance and embodied knowledge, leading to a shift in consciousness and conscience (Pruzan 2014).

The corresponding challenges to our academic institutions are imposing: (i) To develop research and teaching environments that provide a perspective on leadership as an existential search for core-meaning and integrity, whereby a leader's thoughts, words and deeds are in harmony, and (ii) To motivate and communicate such a spiritual-based perspective to leaders who may consider it to be foreign to the context they work in, often dominated by economic rationality and an emphasis on short-term results.

In other words, our institutions of higher learning will face the challenge of integrating spiritual-based leadership in their educational programs without pragmatically reducing it to an instrument of economic rationality—of developing and mediating a leadership paradigm that cannot be taught but which must be accessed via the emancipation of embodied knowledge (Pruzan 2014).

References

Pruzan, P. 2009. *Rational, ethical and spiritual perspectives on leadership: Selected writings by Peter Pruzan*. Oxford: Peter Lang Academic Publishers.

Pruzan, P. 2011. Spirituality as the basis for leadership. In *The Palgrave handbook of spirituality and business,* ed. L. Bouckaert and L. Zsolnai, 287–295. London: Palgrave.

Pruzan, P. 2014. The source of ethical competency: Eastern perspectives provided by a Westerner. In *Business and the greater good—Rethinking business ethics in the age of economic crisis,* ed. Knut J. Ims and Lars Jacob Pedersen. Cheltenham: Edward Elgar Publishers (forthcoming).

Pruzan, P., and K. Pruzan Mikkelsen. 2007. *Leading with wisdom: Spiritual-based leadership in business*. New Delhi: Sage and Sheffield, UK: Greenleaf Publishing.

Pruzan, P., and K. Pruzan Mikkelsen. 2010. *Ledelse med visdom: Spirituelt baseret lederskab i virksomheder* (revised and translated version of "Leading with Wisdom: Spiritual-based Leadership in Business"). Copenhagen: Gyldendal Business.

Peter Pruzan is professor emeritus at the Copenhagen Business School (CBS). He has degrees from Princeton University (B.Sc.), Harvard University (MBA), Case-Western Reserve University (Ph.D.) and the University of Copenhagen (Dr. Polit.). He is Visiting Professor at Sri Sathya Sai Institute of Higher Learning in India. Among his organizational initiatives are: co-founding, designing or governing: Bachelor's and master's programmes at CBS in Mathematics and Economics (1987) and in Philosophy and Economics (1996); The Institute of Social and Ethical AccountAbility in London (now: AccountAbility), London, 1996; The Copenhagen Centre—New Partnerships for Social Responsibility, Copenhagen, 1998; CBS Centre for Corporate Social Responsibility, Copenhagen, 2002; European Academy for Business in Society, Brussels, 2002. His most recent books are "Rational, Ethical and Spiritual Perspectives on Leadership: Selected Writings by Peter Pruzan" (2009) and, together with his wife Kirsten, "Leading with Wisdom: Spiritual-based Leadership in Business" (2007/2010).

Educating Moral Business Leaders Without the Fluff and Fuzz

Joanne B. Ciulla

Peter Pruzan defines spiritual-based leadership as "an inclusive and yet highly personal approach to leadership that integrates a leader's *inner* perspectives on identity, purpose, responsibility and success with her or his decisions and actions in the *outer* world of business." He argues, "spiritual-based leadership implies not just a transformation of the teaching and the practice of leadership, but also, and more fundamentally, the transformation of the individual leader's mind-set." I do not take issue with Pruzan's idea of spiritual-based leadership for the reason that Pruzan says academics object to it—because it is 'fluffy' and cannot be empirically tested. The problem, as I see it, is not that Pruzan's idea of spiritual leadership is 'fluffy' but that it is 'fuzzy'. One cannot tell from his description what spiritual leadership means, how differs from ideas in a variety of other ideas in philosophy and leadership studies. In short, the problems with spiritual-based leadership are primarily conceptual not as Pruzan says, pragmatic. You cannot teach or have spiritual-based leadership unless you understand what it is and what it does. However Pruzan argues that making it 'doable' "may lead to its self-eradication." This papers endeavors to suggest other ways of articulating what Pruzan wants to say that are doable for those who teach and practice leadership.

Let us begin with the description of a spiritual-based leader. Pruzan says that spiritual leaders are "nourished by their spirituality and that their spirituality guides them." There seems to be a leap from inner spirituality to the outer world and from self-fulfillment to the fulfillment of others. What makes this leap happen? How does spirituality "integrate" the outer and inner world and in particular, how does it connect leaders to followers and other stakeholders.

When we look at the bad behavior of leaders in business and other sectors, we should ask, "What is the problem that we are trying to solve?" Is it that they lack spirituality? Again, that depends on what spirituality means. Does spiritual mean ethical? Leaders face two broad categories of ethical challenges. The first

J. B. Ciulla (✉)
University of Richmond, Richmond, USA
e-mail: jciulla@richmond.edu

L. Zsolnai (ed.), *The Spiritual Dimension of Business Ethics and Sustainability Management,* DOI 10.1007/978-3-319-11677-8_15

is personal and has to do with the leader. They sometimes lack concern for other humans and living things. We also know that they may see themselves as 'special' or guided by greed or the desire to "win" in a competitive market. The second, problem is cognitive or epistemological. It has to do with how they think, what they know and what they think constitutes truth about the world and the way it works. This would include things like their knowledge and beliefs about the market system and the notion that the primary truths are those that come from numbers.

Several philosophers and researchers touch on what Pruzan says about the consciousness and personal character of the leader. Confucius observes, "If a ruler sets himself right, he will be followed without his command. If he does not set himself right, even his commands will not be obeyed" (Confucius, Analects 13.6). Spirituality may be a way of 'setting oneself right' but the humanistic approach emphasizes doing so through others. Confucius writes, "A man of humanity, wishing to establish his own character, also establishes the character of others, and wishing to be prominent himself, also helps others to be prominent" (Confucius, Analects 6.28). Empirical studies of authentic leadership focus on self-knowledge. Bruce Avolio and Fred Luthans define authentic leadership as "a process that draws from both positive psychological capacities and a highly developed organizational context, which results in both greater self-awareness and self-regulated positive behaviors on the part of leaders and associates, fostering positive self-development" (Luthans and Avolio 2003, p. 243). This definition later adds explicit and 'doable' moral elements such as: "a pattern of leader behavior that draws upon and promotes both positive psychological capacities and a positive ethical climate, to foster greater self-awareness, an internalized moral perspective, balanced processing of information, and relational transparency on the part of leaders working with followers, fostering positive self-development" (Walumbwa et al. 2008, p. 94). Authentic leadership is basically about how a leader's self-knowledge contributes to making him or her an effective and a moral leader. There is an inherent circularity in this definition of authentic leadership. Morality seems to be both the cause, result, and quality of being authentic. One senses a similar problem with spiritual-based leadership.

Spiritual-based leadership seems to entail a more introspective process than servant leadership in which leaders grow and develop by improving the people that they serve (Greenleaf 1977). Spirituality may not be necessary for ethical leadership; however, Pruzan believes that spirituality motivates leaders to be ethical. Yet, people can also motivate each other to be ethical. The Southern African concept of *ubuntu* offers another example of how identity and moral growth are connected to the community. Archbishop Desmond Tutu says that a person with ubuntu is open and not threatened by others because "he or she has a proper self-assurance that comes from knowing that he or she belongs in a greater whole and is diminished when others are humiliated or diminished, when others are tortured or oppressed, or treated as if they are less than who they are" (Tutu 2000, p. 31).

Philosophers such as Confucius and ancient Greeks such as Socrates, Thucydides, and Protagoras refer to the virtue of reverence in a way that seems similar to spirituality. The ancient Greeks emphasize reverence as the primary virtue for leaders. As philosopher and classicist Paul Woodruff writes in his book *Reverence: Renewing a Forgotten Virtue*, "reverence is the virtue that keeps leaders from trying to take tight control of other people's lives. Simply put, reverence is the virtue that keeps human beings from trying to act like gods" (Woodruff 2001, p. 4). According to Woodruff, reverence consists of the ability to feel awe and a profound respect for others that comes from the realization that leaders and followers are all part of a larger whole. Reverence is a kind of umbrella virtue that encompasses humility, respect for persons and the world around us, and a sense of awe about things greater than the self. Today people usually think of reverence in relation to religion however, it is also a secular notion that makes an explicit connection between the self and others, leaders and followers. Pruzan touches on this idea when he talks about the "feeling of unity" that comes with love. Since reverence is a virtue, which is not a feeling but a practice and way of doing things, it is easier to get around Pruzan's "paradox of pragmatism." Aristotle defines virtues as conscious habits or things that we do. We learn them through education and role models (Aristotle, Nichomachean Ethics, Bk. 2.5).

Whether one talks about spirituality or virtues, educators face the challenge of how to best develop ethical and effective leaders (Ciulla 2011). In the 30 years that I have worked in business ethics, I have been delighted to see how the field grew and developed around the world. Nonetheless, it pains me that the battle to teach ethics courses in business schools continues, not with the business community but with business schools. Many of them are still not interested in investing in business ethics faculty and courses because, despite scandals and the crash of the global financial system, they still do not think that business ethics is important.

Like Pruzan, I was concerned when CSR began to emerge in Europe. While it encompasses important aspects of business ethics such as sustainability, CSR focuses more on the behavior of the firm and than on the personal development of students as future leaders. In response to this trend, my friend the late Robert C. Solomon and I decided to publish a textbook that presents business ethics to students in terms of "the good life" (Ciulla et al. 2007). The book includes a long section on the nature of the good life, which is cast in Aristotelian terms of morality, society, and happiness as flourishing (*eudemonia*). We address students in the second person in the book and ask how their decisions and positions affect their ability and the ability of others to lead the good life.

Spiritual-based leadership is only one of many ways to describe how to develop and educate moral leaders. Pragmatism should not be a problem or "paradox" but part of a solution. We want business leaders who do good things. Whether they are motivated by spirituality, humanism, virtue, or an enhanced intellectual perspective is not as important as whether they behave ethically. As educators and scholars we must seek the most effective ways to educate business students to take on the moral responsibilities of leadership but we can only do this if business schools let us. So while I am sympathetic to the ideals of spiritual-based leadership, I doubt that spirituality is the best way to convince business schools to change how or what they teach. Descriptions of spirituality are understandably fuzzy (it's the nature of the beast), which explains why some think it is 'fluffy'. The task of educating moral business leaders is too important to approach without clear and compelling ideas about what moral leadership is and how to practice it. If we get rid of the fuzz, people may stop calling ethics fluff.

References

Aristotle. 1985. *Nichomachean ethics*. (edited by T. Irwin). Indianapolis: Hackett Publishing Co.

Ciulla, J. B. 2011. Ethics and effectiveness: The nature of good leadership. In *The nature of leadership*, 2nd ed., ed. J. Antonakis and D. Day, 508–540. Thousand Oaks: Sage.

Ciulla, J. B., C. Martian, and R. C. Solomon. 2007. *Honest work: A business ethics reader*. New York: Oxford University Press.

Confucius analects. 1969. In *A source book in Chinese philosophy: Selections from the Analects* ed. Wing-Tsit Chan, 14–49. Princeton: Princeton University Press.

Greenleaf, R. 1977. *Servant leadership: A journey into the nature of legitimate power and greatness*. Mahwah: Paulist Press.

Luthans, F., and B. J. Avolio. 2003. Authentic leadership development. In *Positive organizational scholarship*, ed. K. S. Cameron, J. E. Dutton, and R. E. Quinn. San Francisco: Berrett-Koehler.

Tutu, D. 2000. *No future without forgiveness*. New York: Image Doubleday.

Walumbwa, F. O., B. J. Avolio, W. L. Gardner, T. S. Wernsing, and S. J. Peterson. 2008. Authentic leadership: Development and validation of a theory based measure. *Journal of Management* 34 (1): 89–126

Woodruff, P. 2001. *Reverence: Renewing a forgotten virtue*. New York: Oxford University Press.

Joanne B. Ciulla is Professor and Coston Family Chair in Leadership and Ethics at the Jepson School of Leadership Studies, University of Richmond, USA. She has a B.A., M.A. and Ph.D. in philosophy and has held the UNESCO Chair in Leadership Studies at the U.N. International Leadership Academy in Jordan and academic appointments at La Salle University, Harvard Business School, and The Wharton School. She is the author of numerous books and articles on leadership ethics and business ethics. Ciulla recently co-edited a 3-volume set Leadership Ethics (Sage 2013). She is president of the International Society for Business, Economic and Ethics.

In Admiration of Peter Pruzan's Proposal for Spiritual-based Leadership

Stephen B. Young

Peter Pruzen presents a vital and viable concept—spiritual based leadership—but with unnecessary diffidence. Where could any genuine leadership come from if not from something spiritual,—some sense of the "ought-to-be" presented in opposition to "things-as-they-are"?

Leadership resonates with vision and mission, looking beyond the present, choosing one course over another (Selznik 1957).

But Pruzen, with excessive deference to some preferences of the contemporary academy, notes with concern objections to a formal concept of spiritual based leadership that such concept does not lend itself easily to "rigorous empirical research" and so to data-driven assessments of its operational impacts in the worldly realm of "facticity" (Habermas 1996).

Apropos of privileging empirical fact-finding as the sine-qua-non of intellectuality, was it not Einstein who said: "Everything that can be counted does not necessarily count; everything that counts cannot necessarily be counted".

Presenting leadership as a social fact within human communities of mutuality as the outcome of what Pruzan calls "existential search for core-meaning and integrity" is most practical given the intellectual endeavors of cultural anthropology and social psychology. Parts of academia are well-positioned to explore and explicate the phenomenon of leadership. The thick description of anthropology, for example, can be a starting point for rigorous empirical research into leadership dynamics.

Pruzan's concept of "spiritual-based leadership" is not far from Weber's presentation of charisma as the basis for authority and followership by others (Weber 1947, p. 358). What causes others to follow and to acknowledge one's authority is indeed leadership. Charismatic leadership is personal and not positional. Lack of personal charisma exposes many in positional hierarchies as not deserving of their asserted cultural, social, and political superiority and so to be in want of true leadership abilities.

S. B. Young (✉)
The Caux Round Table, Minneapolis, USA
e-mail: steve@cauxroundtable.net

L. Zsolnai (ed.), *The Spiritual Dimension of Business Ethics and Sustainability Management,* DOI 10.1007/978-3-319-11677-8_16

Where there is no charisma indeed is just where command and control becomes extremely instrumental and manipulative of fear and greed, customary prejudices of inferiority, and pre-dispositions to submission and obedience.

But charismatic individuals can be found and described. Thick description of their dynamics, of how they are regarded and perceived by those who award them with charisma, of what affect they induce in followers, can be recorded to serve as relevant data on human prospects for action.

Erik Erikson's studies of personal identity provide valuable insights into how people develop a charism which qualifies them for leadership. His biographies of Martin Luther and Mohandas Gandhi are two samples of his approach (Erikson 1958, 1969).

In this connection I am reminded of David Riesman's work on the inner-directed and other-directed personalities (Riesman 1950). The inner-directed person has a core of meaning autonomous and individuated in the psyche. Conversely, the other-director person has no such core and therefore must derive purpose and meaning constantly from others. The other-directed person has no consistency of purpose, is fickle and shiftless in attention and affect, with goals and interests subject to trends in opinion.

Riesman's inner-directed person is more qualified for leadership and his other-directed person is better suited for managerial tasks and followership. The inner-directed personality can act outside the box but the other-directed personality is a creature of organizations and their conformist demands.

A study of inner-direction can give rise to a discourse about spiritual-based leadership, leading to both understanding and improved practice.

Bernard Lonergan's exposition of insight practices provides a different approach to the development of inner-direction through acquisition of truth consciousness (Lonergan 1972, 1992).

Related to the development of inner-direction as a personal capacity is practice in reflective thinking and mindfulness, usually but not exclusively through meditation. And, as this perspective on charisma is pursued, one should not overlook the tradition of virtue ethics in moral philosophy. Virtues—as advocated by Aristotle or Aquinas—partake of some personal capacity for reaching out beyond the immediate and the sensual and so invoke spiritual dimensions of life. But persons with virtue can be described and studied; virtues can similarly be described; and virtuous habits can be acquired.

Confucius also articulated an approach to virtue ethics. The capacities of his "jun zi" or "lordly one" demand on acquisition of some charismatic spiritual core of dedication to ideals and standards which are above the merely ordinary. The way for individuals to seek such excellence and so become leaders is set forth in the *Great Learning* and the *Doctrine of the Mean.*

Second, Pruzan has unnecessary reservations as to the value of placing a burden of instrumentality on abstract concepts so they lose their claim to normative rigor and so become unworthy of moral dignity from compromising with temporal and material short-sightedness.

With data-driven rationality, the enchantment of the world is lost and the human spirit is impoverished. Without enchantment, the inner life of meaning atrophies and people lose the gifts of hopeful and inspiring meaning. Pruzan rightly fears that his standard of spiritual inspiration in the midst of a fallen world would soon be compromised and his dream of enlightened conduct trampled into dust.

Pruzan hopes for a leadership that can critique economic rationality and not subsume itself to base motives and objectives. But Pruzan here sustains ancient dichotomies of rivalry and conflict between self and other, God and Mammon, the ideal and the actual, spirit and matter, truth and subjective perception, self-interest and sacrifice for the common good.

But here again the Weberian concept of charisma can provide reassurance: the charismatic person is in the world but not completely of it. The charism which produces inner-direction and leadership propensity is a link to the transcendent and the sublime. It connects the person in the world with something other-worldly. If the person were to lose that connection, he or she would surrender their charism and become just another administrative supervisor.

To the contrary, Riesman's other-directed person is fully in the world, taking his or her cues as to right and wrong, truth and falsity, from others in society and culture. This person is fully socialized by conventional social expectations and lacks courage to think outside the social box. The other-directed person is a stranger to deep spirituality.

A moral case can be made for operationalizing spiritual-based leadership. Without a deep inner core, the purported leader can easily lose sincerity and fall into hypocrisy. Confucius argued that sincerity was a basic need of the leader (Confucius, Analects 12.10). Without sincerity of commitment, instrumentality must prevail in decision-making.

He affirmed that "High station filled without indulgent generosity; ceremonies performed without reverence; mourning conducted without sorrow—wherewith should I contemplate such ways?" (Confucius, Analects 3.26).

If we lose the spirit, we have only hollow husks of meaning and purpose, of virtue and value. We lose the *Tao*; we become straw dogs. On this point, T.S. Elliot left us his poem *The Hollow Men*:

> We are the hollow men
> We are the hollow men
> Leaning together
> Headpiece filled with straw. Alas!
> Our dried voices, when
> We whisper together
> Are quiet and meaningless
> As wind in dry grass
> Or rat's feet over broken glass
> In our dry cellar
> Shape without form, shade without color,
> Paralysed force, gesture without motion;
> ....
> This is the way the world ends
> Not with a bang but a whimper.

In Christian theology the need is for both works and faith. Faith in God without having works in this world seem an insufficient Christian witness. Qur'an is to the same effect: have faith and do good works.

In his *Julius Ceasar* Shakespeare says of Brutus that "the elements so mixed in him that Nature might stand up and say to all the world 'This was a man'." We are material and spiritual. To insist on one and not the other is to deny the truth of our humanity. Spirit must enter the world through our actions and that process is leadership.

To withdraw from the material, to eschew economic rationality completely is to truncate human experience and flourishing. We are creatures of this world—born to its trials and temptations to finally leave its presence through the finality of death. Jesus was ever so mindful of this, acknowledging in the Lord's Prayer that it was acceptable to ask for our daily bread, but reproving Satan with the naturalism that "man does not live by bread alone but by every word of God."

Confucius said that "To see what is right and not to do it is want of courage. (Confucius, Analects 2.24).

Charisma is of the world just as we are. We bridge normativity with facticity, living in both realms at once. Our highest form of leadership, therefore, should act similarly. It should be at once spiritually transcendent and rationally instrumental.

This is a duality, a dialectic, and not a conceptual absolutism; it is a pluralism and not a fundamentalism, both realistic and idealistic, a stance available to all.

A suitable conceptual framework for this dialect of ought and is, of fact and ideal, is stewardship. Stewardship discourse, standards for stewardship, thick description of good and bad stewardship, can become the apprehended academic discipline of spiritually-based leadership.

The discourse of stewardship can be found in wisdom traditions; in the writings of Cicero on office; in the political thought of John Locke; in the common law of fiduciary duties, even in the paradigm of accountability to others in negligence. The rhetoric of stewardship is that of office, of holding a trust, of using power for the common good.

References

Confucius analects. 1969. In *A source book in Chinese philosophy*: Selections from the Analects, ed. Wing-Tsit Chan. Princeton: Princeton University Press.

Erikson, E. 1958. *Young Man Luther. A study in psychoanalysis and history*. London: Faber and Faber.

Erikson, E. 1969. *Gandhi's truth: On the origin of militant nonviolence*. New York: W.W. Norton.

Habermas, J. 1996. *Between facts and norms*. Cambridge: MIT Press.

Lonergan, B. 1972. *Method in theology*. New York: Seabury.

Lonergan, B. 1992. *Insight: A study of human understanding*. Toronto: University of Toronto Press.

Riesman, D. 1950. *The lonely crowd*. New Haven: Yale University Press.

Selznik, P. 1957. *Leadership in administration*. New York: Harper and Row.

Weber, M. 1947. *The theory of social and economic organization*. New York: Oxford University Press.

Stephen B. Young is the Global Executive Director of the Caux Round Table, an international network of experienced business leaders who advocate a principled approach to global capitalism. Young has published "Moral Capitalism" a well-received book written as a guide to use of the Caux Round Table Principles for Business.Young was born in 1945. He was educated at the International School Bangkok, Harvard College and Harvard Law School. He came to Minnesota in 1981 to be the dean of the Hamline University School of Law. Previously, he had been an Assistant Dean at Harvard Law School.He has also taught at the University of Minnesota Law School, Vietnamese history for the College of Liberal Arts, University of Minnesota and Public Office as a Public Trust for Minnesota State University—Mankato. He has published articles on Chinese jurisprudence, the culture and politics of Vietnam and Thailand, legal education, law firm management, Native American law, the history of negligence, and the law of war.

The Paradox of Pragmatism

Paul de Blot S.J.

The core of this paradox is the epistemological problem of the bridge between internal conceptions and the external face of what we observe; the difference between what we can see and what is seen. This involves questioning what the similarities are between objective reality and subjective imagination. Having a dualistic starting point will not help to bridge the gap between object and subject. A position of unity is the most fundamental and the only possible starting point for dealing with the paradox of dualism, just as we see in the Yin-Yang structure of the Tao. Our world is comprised of a basic unity of two co-components; an external level of 'doing' and an internal level of 'being'; the two faces of one reality which are fundamentally connected, being expressions of the same root.

We face a similar problem of addressing dualism in the physical world. According to the classic laws of thermodynamics, the energy in a closed system is constant because no further energy is available. According to the irreversible entropy law of Clausius, energy irreversibly degrades. Entropy is an indication of the chaotic situation of a system and can never diminish. A chaotic system can never become a harmonious system.

However, in the same physical world another inspired idea about thermodynamics arose and was introduced by Ilya Prigogine, who was influenced by the scientific approach of geologists and biologists. He came to the conclusion that the classic laws of thermodynamics were limited. Living organisms are open systems and are quite different from closed material systems. These open systems continually exchange energy with their environment. Through this process a chaotic external situation is internally expressed through the creation of a harmonious internal structure. In material systems, life arises from inside and expresses itself to the external world (Prigogine 1961, 1977; Prigogine and Nicolis Self 1977).

Paul de Blot S.J. (✉)
Nyenrode Business University, Breukelen, The Netherlands
e-mail: pauldeblot@yahoo.com

L. Zsolnai (ed.), *The Spiritual Dimension of Business Ethics and Sustainability Management*, DOI 10.1007/978-3-319-11677-8_17

According to the classic laws of thermodynamics we live in a dualistic world comprising a closed material system and an open organic reality—two worlds without any intrinsic connection. According to Prigogine, such a conception of dualism is an abstract reflection of a closed system of thinking. In the open system which characterizes our world we may observe living systems with internal harmony existing in an external state of chaos. The 'disturbing' energy on the outside is a creative component of the inner process. Thus the formerly-described 'closed' thermodynamic context can be seen as merely the outside face of an 'open' living system. There is no question about there being a bridge between the systems. Renewed ideas about thermodynamics give us the answer to the problem of dualism: de facto, it does not exist.

Evolutionary theory also faced the same problems with dualism. Descartes created a split between the material and spiritual world and fomented a dualistic world view that has been used in scientific research. The cohesion of body and soul was lost. In evolutionary theory the question of 'the missing link' arose. How did a monkey become a human being? For the well-known specialist in evolution, Teilhard de Chardin, this dualism represented a problem with a mode of thinking which took the evolutionary process to be exclusively developed from the realm of the material world, completely separated from the unseen and immeasurable spiritual dimension (Teilhard de Chardin 1965, 1972).

The existence of a 'bridge' between matter and spirit was thus untenable. Teilhard de Chardin believed that the only way to gain deeper insight into the evolutionary process was to regard evolution as an expression of spiritual unity and see the whole process of evolution as a spiritual process of conscientization of the material world. His basic premise was that there existed a spiritual world (the "milieu divin"; the divine energy environment) which activates spiritual life in the material realm. The material world was a chaotic reality, not a lifeless world, but a spiritual world in *potentia*, not yet conscious of its spiritual life (Teilhard de Chardin 1965, 1972).

Step by step, material non-organic components become active components of life in the organic world of living beings. In the DNA of living beings, material ingredients like iron, copper, oxygen, etc. function as active components of life. In humans these material elements become components of a self-conscious spiritual life system with an inner freedom. Material components become supporting ingredients for our spiritual life. Teilhard de Chardin sees the evolutionary process as a process of the spiritualization of the material world (Teilhard de Chardin 1965, 1972).

A quite different approach to dualism is elaborated by the French existentialist Gabriel Marcel (1949) in reaction to the dualism of Jean-Paul Sartre (1958), also a French existentialist who started from a more dualistic standpoint. Satre states that there is complete separation between our human consciousness and the inanimate world. Humans are condemned to be free and complete freedom is complete isolation. 'Other people' are always the other sides of oneself and thus become one's experience of hell. Sartre's novels and plays describe well the fundamental experience of 'other people as hell' as the price of freedom. If I love somebody, this love can always be lost. A relationship becomes more and more hellish based on my anxiety. The conclusion is to stop having relationships with others to be completely free. Yet making this decision creates another kind of hell. I feel free but I am extremely lonely. I am condemned to be free, but I cannot become a free person if I create a relationship. His notion of Being and Nothingness highlights the quality of his existence as one of "nausea"; a pervasive, even horrible taste of freedom which results from isolating oneself from one's environment; a complete experience of the dualistic split.

Marcel (1949) transcends the closed world of Sartre by going beyond the nothingness with his distinction of "etre et avoir"; having and being. Sartre (1958) isolates himself on the level of having, the level of death and dying. Everything you have, including a relationship, can be lost or can die. There is a difference between having a house and being hospitable. When we hope, we do not *have* hope, but we *are* hope. My body is both something that I have and something that I am. I can *have* "a body" but it *is* not "my" body. As soon as I feel that the body in question is *my* body, this body also is me; it is what I am. In this experience of "my body" I transcend the dualism of *having* in the experience of *being*. The same difference we find in the experience of friendship. "I *have* a friend" is quite different from "he *is* my friend"; my friend is me.

We relate to other things and persons (my body, my friend) using two modes of the same reality; the internal and the external. Having things relates to an external reality; I can possess and dispose of them. Although it is clear that I cannot *have* another person, *being* a friend is meaningful and no-one can take this from me. In taking this non-dualistic attitude 'the other' is heaven. Marcel (1949) also introduced the problem of being through the concepts of having and mystery. A problem is something which I meet, which I find completely before me, external to me. A mystery is something internal; a situation in which I am myself involved. The concept of mystery can only be understood in holistic terms, where the distinction between what is *in* me and what is *before* me loses its duality and becomes meaningful.

The distinction between problem and mystery is one that hinges on the notion of participation, which is also elaborated in metaphysics. This notion of participation can give us deeper insight into the problem of dualism. Thomas Aquinas claims that 'to participate' is to take part in something according to its own nature (Blackbuen 2008). The part is the whole according to the mode of the part (I am Paul as a part of the whole of all beings according to the mode of Paul). Everything IS a part of the totality according to his own individual mode. If participating merely means taking part in something, the participant will not possess the nature of the thing in which it participates in any total fashion, but only in a partial fashion. The participation framework governs what is meant by essence and existence. Essences exist, but they do not exist essentially; they rather participate in their acts of existence. The act of existence is thus limited and thereby individuated to the essence whose act it is.

In summarizing the issue of dualism we must conclude that it represents a fundamentally epistemological problem which comes up in different scientific fields. It relates to the nature of the bridge between the 'outside' and the 'inside' of our experience and cannot be explained by starting from examining the difference between them. We have to start from a position of unity, which is the spiritual inside of our being. Prigogine started by considering an open system with a chaotic outside which develops internal harmony. Teilhard de Chardin considered evolution to be the self-conscientization of the material world which must develop from a spiritual interior. Gabriel Marcel looked for meaningfulness in relationships by reflecting on modes of having and being. In metaphysics and Greek and Scholastic philosophy every being is as a part of the totality of the whole world of beings; the Being.

References

Blackbuen, S., ed. 2008. *Oxford dictionary of philosophy*. Oxford: Oxford University Press.

Marcel, G. 1949. *Being and having*. Westminster: Dacre Press.

Prigogine, I. 1961. *Introduction to thermodynamics of irreversible processes*. New York: Intersciense.

Prigogine, I. 1977. *Time structure and fluctuations*. Stockholm: Nobel Lectures.

Prigogine, I., and G. Nicolis Self. 1977. *Self organization in non-equilibrium systems*. London: Wiley.

Sartre, J. P. 1958. *Being and nothingness. An essay on phenomenological ontology*. London: Matthieu.

Teilhard de Chardin, P. 1965. *The phenomenon of man*. New York: Harper.

Teilhard de Chardin, P. 1972. *Human energy*. New York: Harcourt Brace Jovanovich.

Paul de Blot S.J. was born in Indonesia in 1924. He joined the Allied Forces during World War II and was imprisoned for 5 years. In 1948 he became Jesuit and completed his studies in Indonesia in Indonesian and Javanese language, Eastern cultures, philosophy, politology, physics; and in theology, psychology, philosophy of science and culture, spirituality and business administration in the Netherlands. His philosophy licentiate was about the Eudaimonia of Plato. His PhD was about organizational innovation in a chaotic environment by innovation of men.

His work experience include: In Indonesia chemist in the sugar industry and supervisor of textile and ceramic industry; teaching primary school and high school, teaching theology at the State University in Yogyakarta, philosophy, teaching politology, ecological technology and medical ethics at the Navy Academy. In 2006 he became professor in Business Spirituality at the Business University Nyenrode. In 2010 he got the title "professor honoris causa" there. He is editor of Business Spirituality Magazine.

Reflections on Peter Pruzan's "Spiritual-Based Leadership"

Katalin Illes

A growing area of leadership discourse is based on the spiritual and philosophical dimensions of leader development. The topic is there in academic papers, conference themes and memoirs of some leaders. Nitin Nohria argues that we need leaders who demonstrate moral humility (Starkey 2013). How do you cultivate moral humility? Can it be taught?

Starkey believes that a narrative of common interest is required to combat the mantra of selfishness and to create a sense that leadership is for all and not for the few (Starkey 2013). Can we change the narrative when "almost all scholarship on management and organization is written as if the author is absent, as if objective truth were being recorded?" (Sinclair 2007, p. 444).

A narrative of common interest should have a human voice, a humble and personal voice respecting all other voices.

Great leaders (and followers) quite naturally make the connection between their external and internal worlds and relate to the environment in a humble and responsible manner expressing their concerns for the common good as something deeply personal. Internalizing the concerns of the community enables them to create a bond with others.

To internalize we need to be in silence and listen inwardly. Devoting time to exploring and cultivating the mysteries of one's inner world is an integral part of a spiritual journey. In secular societies contemplation is not a natural part of life. Although meditation and mindfulness training are quite readily available, these practices typically come as an added extra, a novelty rather than an integral part of the community's or individual's life.

Peter Pruzan's article highlights the tension between the external and internal, the material and the spiritual and the role of the academic community. He warns us that by 'dragging' the metaphysical into the physical we lose its essence, its nurturing life force.

K. Illes (✉)
University of Westminster, London, UK
e-mail: k.illes@westminster.ac.uk

L. Zsolnai (ed.), *The Spiritual Dimension of Business Ethics and Sustainability Management*, DOI 10.1007/978-3-319-11677-8_18

I would like to elaborate on two points: firstly, on the dismissiveness of the mainstream academic community towards spiritual based leadership and secondly, on the very personal nature of inner development and the limited yet vital role of a teacher in this process.

Pruzan observes that many in the academic community consider spiritual-based leadership as being "fluffy". In my personal opinion and experience such comments come primarily from those who have not had a chance to explore and develop their inner world yet. Those who try to live their lives only in the external, material world are rather protective of the boundaries of their world view, be that in an academic or business context, and consider all suggestions for an alternative as an attack on their carefully crafted outlook on life. They dismiss or even ridicule the attempts that try to make sense of the world without trying to force everything into measurable, scientific or abstract frameworks.

Most things that are valuable in life cannot meaningfully be counted or measured. Love, friendship, support, trust or a helpful gesture and a smile from a stranger do not have financial value yet life would be unbearably bleak without them. One can only fully appreciate these priceless gifts of life from the inner world. Feelings and emotions are important signals. They reflect our level of well being in any given time. Acknowledging the messages that come from the inside and finding appropriate ways of expressing them are part of one's own journey towards maturity and wholeness. By connecting the signals of our senses and emotions with our cognitive brain we make fuller use of our resources and are better equipped to deal with the growing complexities of our external environment.

If we take trust as an example and talk about it in the abstract it loses almost all of its importance and can easily be reduced to an item on a list. It is easy to talk about trust in general. It is not so easy to be continuously vigilant and observe our own behavior, admitting to ourselves, and when appropriate, to others when we did not act in a fully trustworthy manner. When trust is defined as action and we internally monitor ourselves then trust stays in the forefront of our attention as something fragile and precious that holds human relationships together in a meaningful and fruitful way (Illes 2009).

Humanity is at a cross road and we need to make more responsible choices to face our social, environmental and economic problems. A shift in human consciousness is required. Such a shift cannot happen in the abstract, rhetorically magnified and distorted external world. Such a shift can only take place inside the individual. Those who take personal responsibility for the current state of the world and try to act and behave in a more self-less way are tapping into their spiritual resources and connect with the world internally.

Leo Tolstoy observed that "There can be only one permanent revolution—a moral one; the regeneration of the inner man. How is this revolution to take place? Nobody knows how it will take place in humanity, but every man feels it clearly in himself. And yet in our world everybody thinks of changing humanity, and nobody thinks of changing himself" (Tolstoy 1907, p. 100).

The only way to contribute to the change of humanity is to start the change in ourselves. There is no shortage of knowledge or inspiring examples to evidence this. The wisdom literature from different cultural traditions keep telling us to pay attention to our inner world, to search for the answers in the silence of our hearts, to sit with our dilemmas and patiently wait for the answers, to listen to the inner wisdom that we all carry inside.

Unfortunately intuition and inner knowing are not highly rated in the external world. They are dismissed as intangible, non-rational and non-scientific. Yet, many discoveries and bright ideas surface intuitively in the human brain while the individual is not thinking about the actual problem.

Pruzan's paper inspired me to reflect on my own journey in the context of spiritual-based leadership. I have been teaching leadership for almost 20 years. The titles of my courses and workshops reflected the trends and institutional requirements of the day but the content and the overall message of my teaching has always been spiritual-based and reflected the state of my inner world.

My childhood and upbringing was fortunate partly because it happened in a country where religion was not allowed, spirituality was neither acknowledged nor supported, and only rational and physically measurable activities were valued and rewarded. As a little girl I was eager to meet the external expectations of my restricted world and at the same time I developed a rich inner world without boundaries or limitations.

I did not know then that what is oppressed externally usually finds a way to grow internally. From an early age it was quite natural for me to sit, dream, ask questions in the silence and patiently wait for the answers that usually came through stories and images.

As an adult and an educator I have become passionate about helping people to grow and bring the best out of themselves. It did not take me long to realize that my life purpose [to help others to grow] could not be taught from textbooks and I would not be able to achieve anything meaningful with others unless I focused on my own development. I noticed that as my inner world expands my external world reflects the changes and mirrors my spiritual landscape.

Reviewing my actions regularly enable me to become more forgiving and less judgmental towards myself and others.

In my classes I try to create a safe learning environment that supports both the external and internal processes of all participants. Each event is unique and the outcomes are unpredictable. Knowing that I am only one of the components in a complex process makes me very humble and vigilant. Making myself vulnerable by showing parts of my inner landscape, when appropriate, enhances my human connection with people and enables me to form an invisible yet almost tangible bond of trust and openness.

Peter Pruzan's article gave me an opportunity to reflect on the values of spiritual based leadership. Now I appreciate even more the interconnectedness between my inner world and how my embodied knowledge is mirrored through the choices I make, the relationships I build and the impact I have on the world.

I believe that connecting with people whose lives are informed by spirituality is a valuable way to grow. Teachers and leaders who are already on the spiritual path have a responsibility to be there for others and hold a safety net for those who would like to start exploring the mysteries of their inner world.

References

Illes, K. 2009. Defining trust as action: An example from Hungary. *Philosophy of Management* 7 (3): 20–27.

Sinclair, A. 2007. *Leadership for the disillusioned*. London: Allen & Unwin.

Starkey, K. 2013. Academic view: A new philosophy of leadership. *The Economist Online*. http://www.economist.com/whichmba/academic-view-new-philosophy-of-leadership. Accessed 11. Dec 2013

Tolstoy, L. 1907. Three methods of reform. In ed. *Pamphlets,* Leo Tolstoy, London: A.C. Fifield.

Katalin Illes is a Principal Lecturer in Leadership and Development at the University of Westminster in London. She has a wealth of experience in teaching Leadership, Personal Development, Management Development and Intercultural Communication on post-graduate and executive programmes. She is a passionate promoter of transformational change through leadership development and has extensive international work experience. She held strategic leadership positions in the UK and China. Katalin has also contributed to collaborative partnership developments in India, Hungary and Switzerland. Katalin Illes's current research interests include ethical leadership, building trusting relationships in organisations and innovative ways of leader development.

Further Reflections on Spirituality and Spiritual-based Leadership

Peter Pruzan

I am deeply impressed by and thankful for the reflections I received from my four respondents. In spite of the significant differences in their perspectives and attitudes, together they provide considerable food for thought as to the nature of the challenges raised in my brief paper. Those challenges dealt with developing research and teaching environments that foster a perspective on leadership as a spiritual search; motivating and communicating such a perspective in environments dominated by economic rationality; and integrating spiritual-based leadership into educational programmes without reducing it to an instrument of economic rationality.

That the responses to my paper and the challenges differ considerably in their foci is perhaps not surprising given the nature of the topic; according to Joanne Ciulla, "Descriptions of spirituality are understandably fuzzy (it's the nature of the beast)…".

Furthermore, while it originally surprised me, I now appreciate the emphasis that, in particular Paul de Blot, places on the concept of duality. I presume this was triggered by my opening statement that "spiritual-based leadership is emerging as an inclusive and yet highly personal approach to leadership that integrates a leader's *inner* perspectives on identity, purpose, responsibility and success with her or his decisions and actions in the *outer* world of business…" as well as by my reference to "external actions" and 'internal reflections'.

A third and final possible explanation for the considerable difference in the respondents' foci can be found, I believe, in their personal experiences with spirituality as opposed to more intellectual/conceptual/practical orientations. It is my experience that the manner in which people react to spirituality and spiritual-based leadership is highly dependent on whether they themselves are either conscious

P. Pruzan (✉)
Copenhagen Business School, Frederiksberg, Denmark
e-mail: pp.mpp@cbs.dk

Sri Sathya Sai Institute of Higher Learning, Prasanthi Nilayam, Andhra Pradesh, India

L. Zsolnai (ed.), *The Spiritual Dimension of Business Ethics and Sustainability Management,* DOI 10.1007/978-3-319-11677-8_19

seekers on the 'spiritual path' or have had transforming spiritual experiences rather than being primarily academic observers and investigators of leadership as an activity and profession.

In order to deal with these very different responses in a reasonably concise and structured manner, I have chosen to emphasize three themes that they deal with: (a) the nature of spirituality and its relation to leadership (and to ethics), (b) how the propensity of leaders and teachers of leadership to embrace the concept and practice of spiritual-based leadership depends on their awareness of their own spirituality, and (c) the role of dualism with respect to consciousness and spirituality.

1 On Spirituality and Spiritual-Based Leadership

Most of us do not hesitate to employ everyday words (such as stars, biology, ethics, love, life, consciousness…) even though we might find it difficult to define such words. And we naturally assume that others will understand us when we use these words. However, in an academic setting, it is reasonable to expect that the writer has either defined his terms or that the terms are so well-established that a definition is not necessary. This is clearly relevant to my own article on spiritual-based leadership, a concept that is not yet widely used—and judging by the comments I received from the respondents, I feel this applies to several of them as well. Joanne Ciulla in her comprehensive response notes that the concept of spiritual-based leadership is "fuzzy" and that it is difficult to see "how it differs from ideas in a variety of other ideas in philosophy and leadership studies."

So let me expand briefly on this conceptual aspect. I commenced my paper with a brief definition of spiritual-based leadership: "… an inclusive and yet highly personal approach to leadership that integrates a leader's *inner* perspectives on identity, purpose, responsibility and success with her or his decisions and actions in the *outer* world of business…" I developed that definition based on extensive investigation as to how business leaders, who are willing to describe how and why they lead from a spiritual basis, understand spirituality and spiritual-based leadership. For example, the following are how several of the many leaders interviewed for an international research project understand 'spirituality' (Pruzan and Pruzan Mikkelsen 2007; Introduction):

> Knowing the true core of being within you and realizing it is the same core within everyone.
>
> Being so in tune with a universal spirit that you are not acting from a place of ego or desire or greed but are acting on behalf of the welfare of the totality.
>
> Man's quest into his innate divinity—it is more like a road than a state of affairs, a quest more than an arrival.
>
> The manifestation of the perfection that is already there within you.
>
> Our deep connection with a force greater than ourselves—a very individual, lived experience that includes longing and belonging for which the fruits are love and compassion.
>
> The connection among myself, other people and the divine.

I note that it is important to distinguish between spirituality and religion since many theoreticians and practitioners of leadership react strongly to the idea that good leadership is/should be based on a religious orientation (Mitroff and Denton 1999; Mitroff et al. 2009). While religion tends to be characterized by its more formal and institutional aspects (rituals, patterns of worship, dogma, belief systems, priesthoods and sacred texts), the spiritual-based leaders interviewed emphasize that spirituality is characterized by its more introspective qualities. Spirituality, and therefore spiritual-based leadership, entails a personal, existential search for meaning, purpose, truth, identity and interconnectedness that may or may not include a focus on a divine/transcendent/god. This can be seen in the distinctions made by several of the business leaders interviewed in the fore-mentioned research project (Pruzan and Pruzan Mikkelsen 2007; Introduction):

> Spirituality is about how you live your entire life. I see religion as something that has been handed down to you. You may choose to apply some of its values in your life, and your life might become more spiritual because of that. But religion is something external to you whereas spirituality is something that is within you.
>
> Spirituality and religion are not the same. Religion encompasses the belief system associated with a world view, patterns of worship and ritual prayer. Spirituality in contrast is specific to the individual and encompasses each person's lived experience.
>
> While religion offers many beautiful things like rituals and ceremonies, to me it is not spirituality. Spirituality has no borders or restraints. It doesn't separate, it connects.

A couple of the responses to my paper also indicate that there is a lack of clarity as to the distinction between spirituality and ethics. Ethics is generally regarded as a set of principles or rules of conduct that guide individual or collective behaviour. The internationally renowned researcher and teacher of business ethics, Ciulla argues in her response that "We want business leaders who do good things. Whether they are motivated by spirituality, humanism, virtue or an enhanced intellectual perspective is not as important as whether they behave ethically." Yes indeed, we want "business leaders who do good things", but from my perspective, she puts the cart before the horse; spirituality is the well-spring, the fundamental source of true enduring ethical behaviour. The primary source of human ethical competency and our promptings to act in accord with this competency is our inherent but not always readily accessible divine nature, not external stimuli in the form of tradition and moral prescriptions (Pruzan 2014). I realize that such metaphysical presuppositions are not yet 'mainstream' and are not susceptible to traditional scientific investigation—which is a major reason that I focused on the paradox of pragmatism in my paper. Which is also the reason that I conclude my paper with the statement: "In other words, our institutions of higher learning will face the challenge of integrating spiritual-based leadership in their educational programs without pragmatically reducing it to an instrument of economic rationality—of developing and mediating a leadership paradigm that cannot be taught but which must be accessed via the emancipation of embodied knowledge."

The embodied knowledge referred to is the knowledge of the Self, the core essence of our being—where such knowledge is in principle available to us all, including of course our leaders, as it is our fundamental nature, but where the skills and the motivation to access such knowledge are often suppressed by the acceptance—and often the idealization—of traditional success criteria emphasizing financial gain. Just like the rest of us, leaders have differing abilities to transcend their egos and to gain access to their true, spiritual nature. This is why I focus so strongly in my paper on the alignment of "a leader's *inner* perspectives on identity, purpose, responsibility and success with her or his decisions and actions in the *outer* world of business."

According to Bouckaert "By reducing ethics to a functional and instrumental management concept we lose something vital. We are crowding out genuine moral feelings and genuine moral commitment, substituting them for rational and technocratic management tools. This substitution fails. Although business ethics took off in the late 1970s and has flourished for two decades, ethics management has failed in recent times to overcome the emergence of new, more sophisticated and hidden forms of opportunism in business." (Bouckaert 2006, p. 193)

In other words, unless a leader's behaviour is grounded on an existential inquiry that leads to self-knowledge and eventually to Self-knowledge, ethics, no matter how it is taught or what ethical codices are developed to guide behaviour in organizations, will, when the chips are down, continue to succumb to the demands of economic rationality, if not earlier. Scandals, large and small, will continue to anger and frustrate us no matter how much focus our business schools place on business ethics and how many Green Papers are developed to promote corporate social responsibility.

2 The propensity of leaders and teachers of leadership to embrace the concept and practice of spiritual-based leadership depends on their awareness of their own spirituality

The second theme in my reflections as to the reactions of the respondents to my paper is the relationship between one's own spiritual awareness and the motivation and ability to promote and practice leadership grounded in one's spirituality.

According to the response of Katalin Illes, "Great leaders (and followers) quite naturally make the connection between their external and internal worlds and relate to the environment in a humble and responsible manner expressing their concerns for the common good as something deeply personal. Internalizing the concerns of the community enables them to create a bond with others." Later on, she writes about "… the very personal nature of inner development and the limited yet vital role of a teacher in the process." Still later she tells us: "A shift in human consciousness is required. … Those who take personal responsibility for the current state of the world and try to act and behave in a more self-less way are tapping into their spiritual resources and connect with the world internally." These are statements that could only be made by a person who herself is intimately guided by her own spirituality—and who therefore is motivated to 'teach what cannot be taught'.

Illes also provides advice as to how such motivation can be operationalized: "The only way to contribute to the change of humanity is to start the change in ourselves. There is no shortage of knowledge or inspiring examples to evidence this. … As an adult and educator I have become passionate about helping people to grow and bring out the best in themselves. It did not take me long to realize that my life purpose (to help others to grow) could not be taught from textbooks and I would not be able to achieve anything meaningful with others unless I focused on my own development. I noticed that as my inner world expands my external world reflects the changes and mirrors my spiritual landscape."

Similar reflections are provided in the response of Stephen B. Young. He commences with the question: "Where could any genuine leadership come from if not from something spiritual…?" and then relates such a spiritual-based leadership to Weber's concept of charisma: "Charismatic leadership is personal and not positional. … When there is no charisma indeed is just where command and control becomes extremely instrumental and manipulative of fear and greed, customary prejudices of inferiority, and pre-dispositions to submission and obedience. … the charismatic person is in the world but not completely of it. The charisma which produces inner-direction and leadership propensity is a link to the transcendent and the sublime." Towards the end of his response Young states: "A moral case can be made for operationalizing spiritual-based leadership. Without a deep inner core, the purported leader can easily lose sincerity and fall into hypocrisy. … Without sincerity of commitment, instrumentality must prevail in decision-making." As was the case with Katalin Illes, such powerful personal statements and sentiments are expressions of Young's own spirituality.

As did Illes, Young also argues that the teaching of spiritual-based leadership can be facilitated via the identification and description of role models: "… charismatic individuals can be found and described. Thick descriptions of their dynamics, of how they are regarded and perceived by those who award them with charisma, of what affect they induce in the followers, can be recorded to serve as relevant data on human prospects for action." Similarly he refers to the tradition of virtue ethics in moral philosophy and argues that "persons with virtue can be described and studied; virtues can similarly be described; and virtuous habits can be acquired." Such a pedagogical method most certainly is at the heart of communicating about spiritual-based leadership and is a path that my wife and I followed when, together with an American couple, Debra and William Miller, we worked on the spiritual-based research project that led to the book *Leading with Wisdom* (Pruzan and Pruzan Mikkelsen 2007).

It is my conviction that the ability to embrace, develop and promote the practice of spiritual-based leadership depends on one's own spirituality—which is why it is important for both theoreticians and practitioners of leadership to inspire others to establish contact with their own spirituality—with their own "true core", "innate divinity", "perfection that is already there within you", "deep connection with a force greater than ourselves".

3 Dualism, Consciousness and Spirituality

In particular Paul de Blot bases his reflections on the concept of dualism; in fact his entire paper focuses on this concept as one meets it in various domains. He commences with the statement: "The core of this paradox (he refers to the title of my paper: Spiritual-based Leadership: A Paradox of Pragmatism) is the epistemological problem of the bridge between the inside and outside of our observation, the difference of what we can see and what is seen." And he then states: "Unity is the most fundamental starting and the only possible starting point for solving dualism … Our world is supported by a basic unity with two co-components, an outside level of doing and an inside level of being …" He concludes with the statement: "We have to start from the unity, which is the spiritual inside of our being."

Inspired by his reflections I will now conclude my own reflections by briefly considering the classic Cartesian dualism between matter and mind as this is at the heart of any deeper discussion of spirituality. The question I will focus on is: Is consciousness a product of matter or is it itself a fundamental aspect of sentience? In other words, is consciousness caused or is it causal? (Kelly et al. 2007; Chap. 2)

First a few words on 'consciousness', something that we all are conscious of (!) while we know little about its source—and this applies as well to scientists and philosophers. We are unable to answer questions such as: Where do thoughts come from? Where do morals come from? Preferences? Love? Conscience? Self-awareness? Yet we are all able to think, to distinguish between good and bad, to have preferences …. According to (Gurcharan 2009, pp. 148–149) such questions reduce to: "How can we provide objective, third person accounts of what is essentially a subjective first person phenomenon…?"

Descartes formulated an ontological distinction, often referred to in philosophy as the 'mind-body problem', the distinction between the mind as non-physical pure thought, although related to the brain, and the body/world as extended and material: "… it is certain that this 'I' (- that is to say, my soul, by virtue of which I am what I am -) is entirely (and truly) distinct from my body and that it can (be or) exist without it." (Descartes 1641; Meditation VI, pp. 69–70). In spite of this distinction, Descartes argues that the non-extended immaterial mind and the extended material body causally interact. In other words, a central theme in what is often referred to as Cartesian dualism is that while they are ontologically distinct, the immaterial mind can cause material events and the material body can cause mental events.

Cartesian dualism, and with it the concept of a soul or Self, has been discarded by most Western philosophical and scientific thought. Since the Enlightenment there has been a rather broad consensus among such domains to reject this dualism and to accept instead a monist materialist position whereby consciousness is essentially material. Such a position is e.g. at the heart of the cognitive and behavioural neurosciences that thereby ignore the possibility of consciousness as having a non-physical origin and are not able to include concepts of spirit, Self and transcendence.

Yet there are an increasing number of scientists who reject the materialist position. For example, according to the physicist Peter Russell "How can something as immaterial as consciousness ever arise from something as unconscious as matter … rather than trying to explain consciousness in terms of the material world, we should be developing a new worldview in which consciousness is a fundamental component of reality." (Russell 2003, pp. 28–9).

This perspective is also provided by the Nobel Laureate in Medicine, George Wald (1906–1997), who as a 'hard' scientist found himself compelled by empirical evidence and his own reflections to accept perspectives on science, physical reality and consciousness that "shocked my scientific sensibilities" (Wald 1984, pp. 1–2): "Consciousness seems to me to be wholly impervious to science. It does not lie as an indigestible element within science, but just the opposite: Science is the highly digestible element within consciousness, which includes science as a limited but beautifully definable territory within the much wider reality of whose existence we are conscious. … mind, rather than emerging as a late outgrowth in the evolution of life, has existed always, as the matrix, the source and condition of physical reality—the stuff of which physical reality is composed is mind-stuff. It is mind that has composed a physical Universe that breeds life, and so eventually evolves creatures that know and create: science-, art-, and technology-making animals. In them the universe begins to know itself."

In other words, there are two conflicting metaphysical assumptions regarding the nature of consciousness: The reductionist-materialist position taken by neuroscientists that human consciousness can be reduced to and be explained by the physical properties of our brains, and the spiritual position that although there is an interplay between consciousness and the body whereby consciousness is actualized in the brain via physical processes, it is independent of our physical forms and the interplay terminates when the body dies while the consciousness continues to exist (which is a metaphysical assumption underlying concepts of reincarnation) (Pruzan 2013).

My own experiences leave me with little doubt as to the existence of my Self—and strongly motivate me to embrace the development of spiritual-based leadership.

References

Bouckaert, L. 2006. The ethics management paradox. In *Interdisciplinary yearbook of business ethics,* ed. L. Zsolnai, 191–194. Bern: Peter Lang.

Descartes, R. 1641. *Meditations*. (trans. from the Latin with an introduction by L. Lafleur, 1951). New York: The Liberal Arts Press.

Gurcharan, D. 2009. *The difficulty of being good: On the subtle art of Dharma*. New Dehli: Penguin Books.

Kelly, E. F., E. W. Kelly, A. Crabtree, A. Gauld , M. Grosso, and B. Greyson. 2007. *Irreducible mind: Toward a psychology for the 21st century*. Lanham: Rowman & Littlefield.

Mitroff, I. I., and E. A. Denton 1999. *A spiritual audit of corporate America: A hard look at spirituality, religion, and values in the workplace*. San Francisco: Jossey Bass.

Mitroff, I. I., E. A. Denton, and C.M. Alpaslan. 2009. A spiritual audit of corporate America: Ten years later (Spirituality and attachment theory, an interim report). *Journal of Management, Spirituality and Religion* 6 (1): 27–41.

Pruzan, P. 2013. *Perspectives on research methodology for reflective scientists*. Prasanthi Nilayam: Sri Sathya Sai Institute of Higher Learning.

Pruzan, P. 2014. The source of ethical competency: Eastern perspectives provided by a Westerner. In *Business and the greater good—Rethinking business ethics in the age of economic crisis,* ed. Knut J. Ims and Lars Jacob Pedersen. Cheltenham: Edward Elgar Publishers. (forthcoming)

Pruzan, P., and K. Pruzan Mikkelsen. 2007. *Leading with Wisdom: Spiritual-based Leadership in Business*. New Delhi: Sage (for South Asia only) and Sheffield, UK: Greenleaf Publishing.

Russell, P. 2003. *From Science to god: The mystery of consciousness and the meaning of light.* Novato: New World Library.

Wald, G. 1984. Life and mind in the universe. *International Journal of Quantum Chemistry; Proceedings of International Symposium on Quantum Biology and Quantum Pharmacology* 26 (11): 1–15.

Peter Pruzan is professor emeritus at the Copenhagen Business School (CBS). He has degrees from Princeton University (B.Sc.), Harvard University (MBA), Case-Western Reserve University (Ph.D.) and the University of Copenhagen (Dr. Polit.). He is Visiting Professor at Sri Sathya Sai Institute of Higher Learning in India. Among his organizational initiatives are: co-founding, designing or governing: Bachelor's and Master's programmes at CBS in Mathematics and Economics (1987) and in Philosophy and Economics (1996); The Institute of Social and Ethical Accountability in London (now: Accountability), London, 1996; The Copenhagen Centre—New Partnerships for Social Responsibility, Copenhagen, 1998; CBS Centre for Corporate Social Responsibility, Copenhagen, 2002; European Academy for Business in Society, Brussels, 2002. His most recent books are "Rational, Ethical and Spiritual Perspectives on Leadership: Selected Writings by Peter Pruzan" (2009) and, together with his wife Kirsten, "Leading with Wisdom: Spiritual-based Leadership in Business" (2007/2010).

Part V
Conclusions

Spirituality for Business Ethics and Sustainability Management

László Zsolnai

This concluding chapter summarizes the main messages from the book about the promotion of the development of business ethics and sustainability management. One set of messages relates to *research* and addresses how it is possible to incorporate spirituality into business ethics and sustainability management theory. The other set of messages concentrates on *practice* and seeks to find answers to how new working models of ethical business and sustainable economies can be developed and implemented, and how leadership can be transformed to enable ethical and sustainable change.

1 New Insights in Business Ethics and Sustainability

Luk Bouckaert forcefully argues that a spiritual approach to business ethics is badly needed. Without greater intrinsic motivation, business ethics will be reduced to an instrument for reputation and risk management and any genuine moral commitment will be lost. The paradox of business ethics always emerges when ethics is not practiced for its own sake but when it is used to gain material benefit.

The present economic and financial crisis highlights the *paradox of business ethics*: more ethics management does not imply a consistent and integral commitment to ethics when ethics is applied in a selective and market-driven way. Bouckaert shows how, more than with previous scandals, the current financial and debt crisis has revealed the limits of business ethics as a practice of moral self-regulation. Although many of the banks involved in the crisis were committed to CSR programs and had started ethical investment funds, this did not help them to anticipate and avoid it. One can explain this ethical deficit as a lack of business ethics and see the

L. Zsolnai (✉)
Corvinus University of Budapest, Budapest, Hungary
e-mail: laszlo_zsolnai@interware.hu

L. Zsolnai (ed.), *The Spiritual Dimension of Business Ethics and Sustainability Management,* DOI 10.1007/978-3-319-11677-8_20

remedy as more business ethics and more CSR programs. But this strategy of 'more of the same' will fail if the efforts made with business ethics are not supported by critical reflection about the mechanisms of selective blindness in business ethics.

For the French philosopher *Henri Bergson* (1932) spiritually-driven leaders play a key role in any dynamic conception of ethics. Bergson criticized the classic Kantian and utilitarian theories that claim to have identified fixed norms for good and evil based on universal, unambiguous principles. According to Bergson, these theories overlook the role and example of moral leaders and pioneers. These individuals create new interpretations of value and are motivated by three things: first, a keen sense of social frustration and crisis; second, an intuitive and meta-rational sense of the "élan vital" (the inner dynamic) of history; and third, the ability to speak to and mobilize people.

Bouckaert suggests that Bergson's propositions can create a link between ethics, spirituality and leadership. While rational management may suffice in periods of stability and shared trust, it does not suffice during times of deep and radical change. The distinction between the 'rational manager' and the 'value-driven leader' can gain philosophical depth in the light of Bergson's theory. If we look at the examples of spirit-driven leaders we find that openness of mind is always guided by an *ethic of compassion* which implies a deep sensitivity to the vulnerability of life and to other people.

Emmanuel Levinas and *Hans Jonas* developed a notion of responsibility that does not start from the point of view of universal rights and principles, nor from a conceptual representation of the external world but from the contextual experience of the vulnerability of life. For Levinas (1996) the primary sense of responsibility is awakened by an immediate and non-conceptual experience of the vulnerability of other people. This experience of vulnerability may grow into a universal ethic of compassion and responsibility. Hans Jonas (1984) claimed that a sense of responsibility is generated by a 'heuristic of fear'. Confronted with the planetary impact of modern technology and modern lifestyles we should realize that our planet and the lives of future generations are under threat.

Bouckaert believes that a sustainability defined as 'caring for future generations' illustrates that, on the one hand, a spiritual commitment anticipates every declaration of rights (and makes such declarations possible) but, on the other hand, this spiritual commitment must be made tangible by giving people rights and by transforming the economy according to these rights. Applied to business ethics this means that stakeholder management and business plans must always be preceded by a spiritual commitment to future generations. The awareness that connects us to the common good is the first and most deeply intrinsic incentive which can lead us to set up a social praxis of sustainability in business.

Hendrick Opdebeeck emphasizes that combining entrepreneurship with respect for socio-environmental priorities is not a trivial task. Not only is rationality necessary for this, but also wisdom. Sustainability management questions the basic assumptions of the dominant economic paradigm, such as the need to maximize consumer satisfaction and optimize utility—both of which are grounded on a belief in rational, quantitative progress.

Sustainability management concentrates on economic practices at the micro-level, realizing that on this level one can, through the discovery and exploration of human self-consciousness, combine freedom and order. Consciousness of the self allows man to generate compassion, justice and temperance. Such an outcome is possible only if man's inner or spiritual faculties are not neglected. Sustainability management essentially addresses the individual, small-scale, micro-level. The emergence of concrete sustainable economic alternatives is linked to the development of practical sustainable experiments that stimulate others to turn toward more sustainable practices. The power of example is vital. By extension, and after taking necessary learning cycles into account, such sustainability features can be scaled up to the meso-level (large-scale enterprises with a more human face, for instance)—and macro-level initiatives (like Kyoto and Rio+20).

Opdebeeck states that history has proven that the utopian pre-paradigm offers the opportunity to elaborate crucial aspects of sustainability within a pluralistic context. In the utopian paradigm the essential values required to make sustainability concrete are inspired by *philosophical wisdom* from Greek Antiquity and by *spiritual traditions* like Christianity, Judaism and Islam, not to forget Eastern traditions like Buddhism or Taoism.

Andras Laszlo suggests that turning to examine the *inner perspective* will help achieve a state of sustainability. This is the way of the Sufi, which views life in the light of the essence behind the surface of life. According to Sufism, the human soul is continually seeking to realize its true being and to fulfill its purpose in life. This can be seen not only in each individual's yearning to realize his or her potential to the fullest, but also in our collective unfolding towards the highest expression of human existence. The consciousness of humanity is awakening to the interrelatedness of all life, and this is reflected in a new holistic understanding of the universe.

It is important to make a distinction between innovation and creativity. Changing reality means *innovation.* This requires action by a team, is continuous, takes a long time, and delivers something new to the system. Its impact is measurable and it requires the use of tools and project management. *Creativity*, on the other hand, changes perception and requires thinking and feeling. It represents a challenge to individuals and is discontinuous, taking an instant. Its impact cannot be measured and it requires learning methods fuelled by questions, surprises, and incomplete answers. To be able to create means being able to meet situations in an original and fresh way. Creativity is not about reacting to the environment, or processing it and nor is creativity promoted or even catalyzed by external factors.

Knut Ims investigates the deep ecology approach developed by Norwegian philosopher *Arne Naess* who was inspired by Hinduism, Confucianism and Buddhism on the one hand, and Aristotle, Spinoza and Heidegger on the other. Naess (1989) argues that we need to distinguish between deep and shallow approaches to environmental problems. *Shallow ecology* represents a technocratic attitude to pollution control and resource depletion by supporting rules like 'the polluter pays', and by assuming that symptoms can be treated through technological fixes. Such a shallow approach assumes an anthropocentric worldview in which human affluence is vital. As a consequence, levels and patterns of production and consumption are left unchallenged. In contrast, *deep ecology* assumes a relational, total field perspective that fits into a non-reductionist, non-anthropocentric worldview. The prescribed medicine is to change the basic ideological structure, which ultimately means changing how we as humans regard ourselves. Thus deep ecology redefines the very notion of self as a subject—and opens it up to transformation into an eco-Self.

Næss' view begins with one basic norm: Self-Realization! This is understood as "*Self-realization for all beings!*" The self that should be realized is not the ego self but the larger ecological Self. Næss focuses on the human ability to identify with a larger sense of Self. Humans naturally have this capacity. This can be observed as a cross-cultural phenomenon. Supporting the principles of the peace movement is also a part of Næss' philosophy. But Næss stresses that social justice is not enough. We have to produce and consume less—to tread more lightly and more wisely on the Earth. His motto is "*Simple in means and rich in ends*". This speaks for quality of life instead of standard of living and celebrates the virtues of slowness and smallness in an age of speed and scale.

From a Buddhist point of view *Gabor Kovacs* criticizes Western materialism which has raised human beings 'out' of nature and made them the dominant planetary species. The consequence of having this exclusively anthropocentric worldview is the emergence of a growth-oriented, globalized economy. Consumption-based and greed-driven human lifestyles are resulting in the overuse of the resources of the planet and pollution of the environment to an ever-greater extent. The consequences of these dismal global processes are deforestation, soil degradation, losses of biodiversity, the extinction of species, resource depletion and climate change.

A potential solution is offered by *Buddhist* teachings which deal with suffering and the cessation of suffering. The Buddha's central doctrine, the *Four Noble Truths,* addresses this problem. The First Noble Truth states that the true nature of sentient beings is suffering. The Second Noble Truth asserts that the origin of suffering is widespread unchecked cravings for the ephemeral phenomena of the impermanent world. Misery exists because of our attachment to certain states of impermanent feelings and uncontrolled greed for ephemeral things—especially goods produced by the consumption-based society of our times. The Third Noble Truth declares that suffering can be obviated through the elimination of attachment and greed. This assertion leads to the final truth, The Fourth Noble Truth, which states that the tool for doing this is The Noble Eightfold Path. This is lifestyle advice which refers to 'right behavior' that leads to the ultimate goal of Buddhists; the overcoming of greed, hatred and delusion, the three root causes of suffering.

The foremost Buddhist virtue in regard to the environment is '*non-harming*'. This represents an approach to minimizing impact that prohibits the destruction of life. It does not just refer to the prohibition of certain attitudes, but to the encouragement of active participation in doing good. It means the rejection of violence and the cultivation of the positive values of love and compassion towards all sentient beings. This encompasses preserving biodiversity and being kind to animals. Gratitude towards the non-sentient realm of nature is another important environmental virtue in Buddhism. Forest-groves, trees, mountains, lakes and rivers are objects of reverence for the spirit of gratitude. This infers having the aspiration not to intervene or transform, but to revere and contemplate the intact state of the natural environment.

Kovacs emphasizes that the Buddhist lifestyle is characterized by being content with little, being moderate in daily activity, and practicing simplicity. All of these virtues imply a more sustainable and environmentally-sound livelihood. The realization of these virtues results in the fading of greed and craving and the development of a moderate, frugal way of consuming that has a resounding effect on the quality of nature. In Buddhism, sustainability is not a goal in itself, but a positive byproduct of a virtuous life.

Rita Ghesquière uses examples from literature to demonstrate the connection between sustainability and wisdom. She presents the *fable* as a sustainable good and investigates some fables which foster the idea of sustainability. The fable is a short narrative in prose or verse and contains non-human creatures such as animals, plants or even inanimate objects (mirrors, ink or paper) as protagonists. Fables combine merriness and wisdom. They are fiction but the animal characters act as human beings do. Hence their behavior and their circumstances are experienced by the audience as true-to-life, as an essential part of the experience of life. Anthropomorphism, convincing characters and events and generalizations of practical wisdom are essential components of the fable.

Ghesquière stresses that fables teach us some important lessons. *Labor* (the farmer and his sons), *frugality* (the ants), soberness (the fox) and *productivity* (the lesson of the miser) should come together if we are to overcome the crisis. The steady tenacity of the tortoise will help us to *persevere*. And we should not forget the dream of the moth, which is an excellent guide to *happiness*.

2 Practice and Leadership for Achieving Sustainability

Jean-Paul Close describes a new sustainability venture in Eindhoven, The Netherlands, which takes the form of a new model of a complex society moving toward sustainable progress. He believes that the beginning of the 21st century will enter the history books as a brief but intensely significant period in human history when humankind finally became aware of its spiritual self. This *quantum leap* in human evolution will result in a new, self-aware global human society. Instead of 'having' (possession) and 'taking' (greed) we are learning to 'be' (talent) and to 'give' (create).

The following definition of *sustainable progress* is offered: to work together on an ongoing basis in a healthy, vital, safe and dynamically progressive, self-sufficient human society within the context of the ever-changing natural environment in which we live and act. The complexity of air quality and human health issues provide a direct link to this definition of sustainable progress. One needs to address the four key areas of authority that exist within any society if complexity is going to be tackled. These focal areas are (i) the local population, (ii) the local government,

(iii) business enterprises, and (iv) science and education. None of these authorities normally relate to one other in a purpose-driven way, nor take responsibility for such complex co-creative problems. They all function as parts of a chain of economic dependencies and opportunistic money-driven self-interests, each playing some fragmented, consequence-driven, incomplete role in the complex puzzle. Local AIREAS Eindhoven invited them all to take responsibility together in a new-paradigm venture.

The Local AiREAS Eindhoven experience shows that a new form of society can be quickly developed and implemented to foster true sustainable progress in complex societies using powerful instruments from the old paradigm. This process represents an act of responsibility that challenges old bureaucratic dependencies and structures in a proactive and powerful way. Through voluntarily testing the new society using the stress points of human complexity as it increasingly appears over the coming years, the suffering of major fractions of the human populations may be reduced to the minimum needed for evolution, converting fear into pro-active passion and contributions.

Laurie Michaelis presents the *Quakers Society of Friends*' response to climate change. Climate change can be seen as a problem that has arisen due to overemphasis, both in government policy and in popular culture, on free market capitalism and competitive individualism. This system has supported the innovation that has created technologies and products that have led to unprecedented improvements in the standard of living of many people in industrialized societies. It has also enabled the concentration of wealth which is often linked to political power, or to positions in the social hierarchy. Hence it has also created considerable vested interests in maintaining the status quo.

The *Living Witness Project* supports a number of Quaker groups in their attempts to experiment with collaborative approaches to sustainable living as a response to climate change. Living Witness has provided workshops for about 200 Quaker groups, thereby supporting them to act and live more sustainably. It maintains a network of 500 or more individual Friends and is involving them with local meetings, biannual national gatherings, summer schools and a variety of printed and Internet resources. At local meetings members are encouraged to work with a Sustainability Toolkit and to report on their greenhouse gas emissions and the actions they are taking to reduce them.

British Quakers have come a long way towards developing a coherent approach to sustainability. There is no single shared narrative about what matters, or what should be done about it, yet many Friends are making significant steps towards living more sustainably and they see this goal as part of their shared identity as Quakers. The Quaker approach is grounded in a spirituality that focuses not so much on the human relationship to nature as on individual relationships with others. Most Friends are conscious of being out of tune with the consumer culture that most of us are deeply infected by. This is reflected in the difficulty most of us have with limiting our own consumption. Quakers are one of the groups that might be expected to be most willing and able to move towards sustainable living. The required shift is happening, but it is contested and it seems to many of us to be too slow. Perhaps the main hope for humanity is that many examples both of practical initiatives for sustainable living, and of the spiritual and social practices that underpin them, are being developed by Quakers and others.

Aloy Soppe, inspired by the economic ideas of philosopher *Rudolf Steiner*, develops propositions to support sustainable finance. Steiner stated that all material objects such as consumer goods, capital goods and services are subject to ageing and therefore intrinsically diminish in quality and value—in contrast to money. Money represents a right to future purchasing power which is of an essentially everlasting quality. This implies that over the course of time money will become 'overvalued' in relation to goods and services. At the same time, interest payments lead to social injustice (in terms of distribution of income) because the quality of material goods and services can only be maintained through labor inputs, while the interest mechanism causes capital to increase in value (sometimes exponentially). This inevitably results in the accumulation of value.

According to Steiner, the accumulated value must be creamed off by means of making gifts from economic actors to the field of culture. If capital is not destroyed voluntarily through gifts from traders then depreciation will occur involuntarily in the form of inflation and bankruptcy. There should therefore be an ongoing system of institutional transfers of value from the economic sector to the cultural sector. Through these gifts capital will acquire a temporary character and will cease to exist if the recipient has consumed it.

Soppe's *first proposition* is that the average interest rate should vary with the rate of growth of the real economy—which is a proxy for organic growth in economic processes. Lower interest rates are not desirable because of their impact on economic opportunities. Higher interest rates should be rejected because of the implicit additional positive time preference that they entail. His *second proposition* concerns the accumulation of wealth. Under the assumption of the existence of a positive interest rate, the absence of a financial tax and the absence of bankruptcies, financial capital grows exponentially and therefore needs to be managed institutionally, in a positive way. A structural imbalance between the monetary sector on the one hand and the real economy on the other leads to the illusion of purchasing power in the hands of the public. The exponential growth of financial capital needs to be managed institutionally if a state of sustainability is to be achieved.

Arundhati Virmani and *François Lépineux* investigate Navdanya as a form of spiritual-based entrepreneurship for an alternative food culture in India. *Navdanya* mobilizes ideas that constitute an ideological frame for political action in India. The movement offers an example of how spiritual values and traditions can be mobilized to create a vision of sustainable agriculture, defend the individual against corporate rights and promote an alternative form of food culture. The multi-level transformational power of Navdanya should also be highlighted: it lays claim to being a catalyst for an economic model based on fairness, a lever for social change and an actor in defense of the common good.

Navdanya's philosophy focuses on peace with nature through *sustainable agriculture* and *organic farming* and *celebrating biodiversity*—symbiotic relationships between plants lead to higher yields—and solidarity economies rather than economies based on competition. It puts forward a holistic view of man's relation to the planet. It affirms that the right to food is a fundamental one, raises awareness of the hazards of using chemical fertilizers and pesticides or genetically modified seeds, and combats bio-piracy and climate change. It defines its goals as organizing and promoting farmers' groups and preserving biological and cultural diversity, as well as battling for seed sovereignty, food sovereignty and water democracy—all themes covered by the slogan 'Earth Democracy'.

Navdanya's philosophy, goals and principles of action have drawn it into partnership with non-Indian movements such as *Slow Food*—an international movement founded in Italy that defends traditional regional cuisine and local economic systems that are committed to creating a future for food and agriculture in which small producers prosper, a culture of ethical consumption develops, and biodiversity and cultural diversity thrives. Navdanya's stand, 'small is beautiful', conforms to the principles of Slow Food. In 2001 Navdanya received the Slow Food Award and became a Slow Food partner—and Navdanya's founder *Vandana Shiva* has served as Vice-President of Slow Food International.

Navdanya can be considered to be a *transformational entrepreneurship*, a form of entrepreneurship that paves the way for—and allows the realization of—the 'great transition' of the earth system that lies ahead, and one which combines the three perspectives of foresight, global civil society and the common good. Transformational entrepreneurship fosters economic, societal and cultural change towards a global civilization that the planet can sustain, and relies on the mobilization of three levers: (i) Foresight—having a vision of the unfolding of the great transition ahead, of a desirable post-transition state for the world, and linking this to the present; (ii) Global civil society—raising the awareness of the general public at all levels, from the local to the global, about the stakes of the great transition; and (iii) The promotion of the common good living systems—emphasizing the importance of preserving and restoring nature, of promoting global justice, of placing human needs at the heart of economic relations.

Janos Vargha describes how the self-interests of nature transformers are enforced using the example of water construction megaprojects. There are many examples of ambitious big dam projects that have transformed the landscape and the environment—profitable for the 'projectors' but disastrous for others such as local people and natural ecosystems.

The arsenal of the *Club of Projectors* consists of exaggerated or false promises, political pressure, and scapegoating. This list can be extended with more items by examining the activities of projectors from antiquity to current times. Projectors (not only Swift's contemporaries but many others from the previous millennia and the following centuries) frequently ignore negative experiences, or at least do not learn from their mistakes. Accordingly, they make the same errors again and again. History is littered with infamous landmarks left behind by the Club of Projectors: ancient irrigation canals in Mesopotamia that salinified soils and made them infertile; the Corinth Canal in Greece, one of Nero's projects that was completed in the nineteenth century but never yielded the promised benefits; the Grand Canal of Alsace that was abandoned at half its planned length because of its negative impacts (but the part which was ultimately built—from which most of the water of the Rhine comes—caused the groundwater table to drop significantly and large wetland forests to perish); the Aral sea that has almost completely dried out as a consequence of the construction of irrigation canals that deprive its two feeder rivers of water. These are just a few examples from a very long list.

A common and distinctive feature of the majority of such projects is that they are linked to the state and to politics. One main reason for this is that, over a certain size—and many of these types of water projects belong to the big, mega or giga category –, such projects require extensive periods of construction, not infrequently decades, and this locks up huge financial resources for long periods of time with very uncertain returns. Another major reason is that the projects usually occupy, influence or reduce in value private and/or state properties, provoking stakeholders to actively or passively resist them. In addition, supporters of hydropower projects must strongly compete against other projectors who offer other types of power plants which are cheaper and require less time for construction. Recently, supporters of such projects have also encountered public opposition in the form of environmental groups as more and more knowledge accumulates about the negative ecological impacts of such projects. Promoters, therefore, desperately seek to justify their public subsidies and fight for support from any kinds of political systems.

Vargha offers a (partial) counterexample of the Gabčíkovo-Nagymaros dam project on the Danube. Environmental groups lead by the Danube Circle managed to stop the construction of the second barrage in 1989 (which had already started at Nagymaros where the site was surrounded by a coffer dam and the river already diverted into a provisional bed). A few years later the coffer-dam was removed and the river was restored to its original bed. In this section the Danube River again flows freely. Here, the Danube Circle lived up to its goal, encapsulated in the motto of the group borrowed from *Comenius*: "Omnia sponte fluent, absit violentia rebus" (Let everything flow free, violence shall be afar from things).

Nel Hofstra focuses on entrepreneurship models inspired by nature. *Eco-innovation* refers to the use of ecological knowledge to bring forth ecological progress. Many firms have been developing ecological innovations; mostly products and processes that reduce environment-related costs. However, eco-efficiency does not guarantee environmental effectiveness. Invention and design with a regenerative character create new opportunities for the practice of eco-innovation. Such products and services recognize the interconnectedness and unity of all life, physical as well as spiritual. They acknowledge Nature as both mentor and teacher because they understand the mutual form of the relationship man has with Nature.

The *cradle-to-cradle* concept promotes the use of waste as a resource for manufacturing in the bio and technosphere. '*Bioneers*' is a movement that suggests novel ways that forces may be joined to confront and mimic design and innovation in a life-conducive way. Regenerative developments are affirmative contributions to living systems (human and biotic) and present entrepreneurial opportunities. They require constant adaptation and the embedding of growth with a deeper meaning and significance for all involved. This way of thinking will open the way to a completely novel realm for entrepreneurs and will simultaneously awaken us to the magnitude of opportunities, instruction and wonder that our Earth can provide.

Peter Pruzan reveals the paradox of pragmatism in *spiritual-based leadership*. Spiritual-based leaders are nourished by their spirituality, which is a source within them that informs and guides them. They search for meaning, purpose and fulfillment in the external world of business and in the internal world of consciousness and conscience. Their external actions and their internal reflections are mutually supportive so that rationality and spirituality may be seen as mutually supportive perspectives.

Pruzan emphasizes that the evolution of spiritual-based leadership implies not just a transformation of the teaching and the practice of leadership, but also, and more fundamentally, the transformation of the *individual leader's mind-set*. What is required is a consciousness that resonates with the conviction that a precondition for the long-term success of purposeful, organized mercantile activity is spiritual-based leadership and not just the pursuit of material gain. We are faced the challenge of developing vocabularies, perspectives and research methods that can support leadership that is spiritually-based. Instead of having a focus on deliberate and willed action that is considered to be the result of logical generalizations and prescriptive principles, this approach implies focusing on the emancipation and empowerment of inner guidance and embodied knowledge, leading to a shift in consciousness and conscience.

Pruzan—responding to comments by *Joanne B. Ciulla*, *Stephen B. Young*, *Paul de Blot, SJ* and *Katalin Illes*—emphasizes that embodied knowledge is the knowledge of the Self, the core essence of our being, where such knowledge resides and is in principle available to all of us, including leaders, as it represents our fundamental nature. However, the skills and the motivation to access such knowledge are often suppressed by the acceptance—and often the idealization—of the traditional criteria of success that place the emphasis on financial gain. This is why we should focus on aligning leaders' *inner* perspectives towards identity, purpose, responsibility and success. This will in turn reflect on their decisions and actions in the *outer* world of business. The propensity of leaders to embrace the concept and practice of spiritual-based leadership depends on their awareness of their own spirituality. So the relationship between one's own spiritual awareness and one's motivation and ability to promote and practice leadership grounded in one's spirituality is crucial.

At the end of this book I would like to refer once again to *Martin Heidegger*. He once wrote that caring for things demands immanence in God. (Heidegger 1985) The book "The Spiritual Dimension of Business Ethics and Sustainability Management" suggests that we too may try to see the *world* as the *face of God* and organize our business accordingly.

References

Bergson, H. 1932. *L'évolution créatrice*. Paris: Presses Universitaires de France.

Heidegger, M. 1985. *Schelling's treatise on the essence of human freedom*. Columbus: Ohio University Press.

Jonas, H. 1984. *The imperative of responsibility: In search of an ethics for the technological age*. Chicago: University of Chicago Press.

Levinas, E. 1996. *Basic philosophical writings*. (Edited by A.T. Peperzak, S. Critchley. and R. Bernasconi). Indianapolis: Indiana University Press.

Naess, A. 1989. *Ecology, community and lifestyle. Outline of an ecosophy*. (Trans. and revised by D. Rothenberg). Cambridge: Cambridge University Press.

Lászlo Zsálnai is professor and director of the Business Ethics Center at the Corvinus University of Budapest. He is chairman of the Business Ethics Faculty Group of the CEMS (Community of European Management Schools–The Global Alliance in Management Education). He serves as editor of the "Frontier of Business Ethics" book series at Peter Lang Publishers in Oxford. With Luk Bouckaert he founded the European SPES Forum in Leuven, Belgium.
Laszlo Zsolnai's most recent books include "Spirituality as a Public Good" (2007. Antwerp, Belgium, Apeldoorn, the Netherlands: Garant), "Frugality: Rebalancing Material and Spiritual Values in Economic Life" (2008. Oxford, UK: Peter Lang Academic Publishers), "Europe-Asia Dialogue on Business Spirituality" (2008. Antwerp, Belgium, Apeldoorn, the Netherlands: Garant), "Responsible Decision Making" (2008. New Brunswick, NJ, London, UK: Transaction Publishers), "The Future International Manager: A Vision of the Roles and Duties of Management" (2009. Houndmills, UK, New York, NY: Palgrave Macmillan), and "The Collaborative Enterprise: Creating Values for a Sustainable World" (2010. Oxford, UK: Peter Lang Academic Publishers). His website: http://laszlo-zsolnai.net.

Index

L. Zsolnai (ed.), *The Spiritual Dimension of Business Ethics and Sustainability Management*, DOI 10.1007/978-3-319-11677-8

CPSIA information can be obtained
at www.ICGtesting.com
Printed in the USA
LVHW041112120420
653141LV00016B/1059

9 783319 116761